negative math

----

negative math

HOW MATHEMATICAL RULES
CAN BE POSITIVELY BENT

An easy introduction to the study of developing
algebraic rules to describe relations among things

ALBERTO A. MARTÍNEZ

PRINCETON UNIVERSITY PRESS

PRINCETON AND OXFORD

Copyright © 2006 by Princeton University Press
Published by Princeton University Press, 41 William Street, Princeton, New Jersey 08540
In the United Kingdom: Princeton University Press, 3 Market Place, Woodstock,
Oxfordshire OX20 1SY

Library of Congress Cataloging-in-Publication Data

Martínez, Alberto A., date.
   Negative math : how mathematical rules can be positively bent / Alberto A. Martínez.
     p. cm.
   "An easy introduction to the study of developing algebraic rules to describe relations
among things."
   Includes bibliographical references and index.
   ISBN-13: 978-0-691-12309-7 (cloth : alk. paper)
   ISBN-10: 0-691-12309-8 (cloth : alk. paper)
   1. Mathematics.   2. Numbers, Negative.   I. Title.

   QA155.M28 2005
   510–dc22   2005043377

British Library Cataloging-in-Publication Data is available

This book has been composed in Sabon with Hel Neue Family Display

Printed on acid-free paper. ∞

pup.princeton.edu

Printed in the United States of America

10 9 8 7 6 5 4 3 2 1

You can use a spoon to drive a screw into a wall. With practice, you can become skillful at it. You can also learn many juggling tricks with the spoon, and thus impress and bewilder people who don't juggle spoons. And you can make all of this more puzzling by calling the spoon a "fork." And you can write books about it and form societies with other people who also juggle spoons called forks. And even then, sure, you can use a spoon to drive screws into a wall.

But a screwdriver is better. And even if you've never seen a screwdriver, you can just as well invent one. It might resemble the spoon in some ways though not in others. So you can keep your spoon as well; for eating soup, for juggling, or even, occasionally, for driving screws into walls. At least until you have more skill with a better tool.

contents

----

figures

----

negative math

chapter 1

Introduction

Most of us are comfortable in the conviction, with Mr. Smith, that

$$2 + 2 = 4.$$

And some of us might be sufficiently unconcerned with math to be amused by people who, by contrast, wonder about two plus two making three, or five, or all numbers at once. But when we need to construct an airplane, or have an employer pay us for hours of work, we have no doubt that two plus two must be four. So we are happy with this simple arithmetical proposition.

But what about some other mathematical propositions? Is it really true that

$$-4 \times -4 = 16?$$

What does this mean physically? Without using this rule, can we possibly build an airplane that will fly safely, by using instead, say,

$$-4 \times -4 = -16,$$

or something else? How close is the connection, really, between physical experience and the standard rules for operations on negative numbers?

Think of your school days. When you were taught basic arithmetic, teachers used many examples to explain problems and operations in terms of relationships among quantities of physical things. The procedures and results usually made sense. The same was true with the introduction to elementary algebra and geometry. But as you advanced to higher courses, gradually you had to learn rules and procedures that were harder to understand in terms of ordinary experience. Despite your confusion, teachers encouraged you to learn the new rules and to apply them consistently, assuring you that the more advanced mathematics was all a necessary consequence of the basic rules. They claimed that it all had very many uses in physics, just like the basic rules of arithmetic. But eventually, if you continued your mathematical studies far enough, you had to accept symbolic rules and propositions without any explanatory reference to physical experience. By then, teachers probably encouraged you to think that mathematics is true irrespective of any correspondence to the physical world. Along the way, if you paid close attention, you may have noticed that not only was the notion of mathematical truth divorced from that of physical experience, but the very notion of truth, in and of itself, was gradually neglected. Instead of regarding results and propositions as true, maybe you learned to characterize them as "correct," "valid," or merely "consistent" with the premises.

Notice the transition. At first, you became convinced of the meaning and truth of mathematical propositions by virtue of practical explanations. But finally you abandoned experience as the justification. Meanwhile, if you also studied physics, it all seemed to depend on mathematics. Hence, you might wonder: how can mathematics be essential for the description of physical phenomena if the rules of mathematics do not need to be based on experience?

The answer lies in the history of mathematics. Even a slight acquaintance with history demonstrates that originally some mathematical rules were based on physical procedures and results, while other rules stemmed from more abstract considerations. Hence, some parts of mathematics correspond to ordinary experience and other parts do not. Some mathematical operations have a close similarity to physical operations, others do not. Accordingly, it is possible to identify those rules that more closely serve to represent observed relations among things. And, by understanding the extent to which other rules do not correspond to observed relations, it is possible to develop new rules that do so to a greater degree.

So we can devise new mathematical rules. Yet how much creative freedom do we have? Can we construct a system in which, say,

$$-4 \times -4 = -16?$$

Actually, yes we can. Mathematical logicians know that one can devise unusual rules of signs and explore their consequences systematically. But this subject is rarely conveyed to general readers. This kind of playful deviance is usually not taught. It's a bit of a secret. We don't want to confuse students who already are busy enough learning minus times minus is plus, along with much else. At universities, once in a while, it is usually allowed that in a backroom basement of a tall ivory tower, some beady-eyed specialist may carry out unnatural experiments with signs. But tentative experiments with the elements of math are not the sort of thing that is usually given press; they are more along the lines of "kids, don't try this at home." The present book, however, shows how to carry out experimental variations of the ways we usually operate with numbers and signs. It illustrates also how to devise such alterations with the aim of describing ordinary perceptions.

Perhaps it is not difficult to convince the novice that we can develop new systems of symbols following specific rules that do not correspond to physical facts. But it is difficult to convince someone that we can modify rules of ordinary numerical algebra to devise new rules that better correspond with matters of everyday experience. To many people it appears impossible to change the "laws of mathematics." But in truth this *can* be done, and it has been done in one way or another, at different points throughout history. Since the mid-1800s, in particular, mathematicians realized clearly that they could invent new geometries and new algebras having principles and results that diverge from those of traditional mathematics. So nowadays there is not just *one* mathematics, really, there are many. There exist a great variety of geometries and algebras, and they are not all equivalent to one another. For instance, some are more useful than others for describing physical processes.

Someone acquainted with such relatively new geometries and algebras might prefer to say that such developments do not constitute alterations in the known laws of mathematics, but instead the creation or discovery of distinct mathematical systems. But nevertheless, it is useful to speak about "changing" the laws of mathematics because it is essentially by modifying established rules that new sorts of mathematics are devised.

Traditional mathematics has been enormously useful and valuable, so we should have good reasons to tamper with it. Granted, one reason is the need to demonstrate that the principles of mathematics are not unique and immutable as they might appear. But the more important reason is to demonstrate that traditional mathematics involves principles that do not correspond clearly or directly to our perceptions. Simple examples can be found in the concepts of number.

Many students of elementary math are puzzled initially by irrational, imaginary, and even negative numbers. But teachers usually convince them to accept such numbers as just as valid as ordinary numbers. Students are taught to not be misled by the names of such numbers into supposing that they are in any way less real or logical than ordinary numbers. Yet there *are* differences. There are many physical situations where such numbers correspond to operations that are simply impossible. For example, given a box with five apples, you cannot physically "subtract," that is, remove, apples in a way that will leave a negative quantity of apples in the box. The physical world prevents you from taking more than those five. Likewise, there is no physical experiment, no measuring device, that when made to obtain a numerical measurement, under whatever circumstance, will yield an imaginary number. The same is true for irrational numbers. For example, no matter what seemingly circular physical object you choose to measure, by trying to fit its diameter several times into its circumference, you *never* obtain π as the result. To be sure, you may obtain a number resembling π, such as 3.14158, but you do not obtain π. This is not to say, of course, that certain machines cannot convey approximations of irrational numbers. Moreover, with or without the aid of calculating machines, such numbers are commonly used in the theories of physical science. They serve to obtain results that do relate to experimental measurements. But in these cases, such numbers are never the results of direct measurements. By contrast, the quantity five can result directly from measurements. Accordingly, we may distinguish how elements of mathematics, such as the different sorts of numbers, correspond either exactly or only approximately to the results of practical operations such as counting and measuring.

The main goals of this book are the following. First, to remind us that some aspects of traditional numerical algebra do not correspond to our everyday experiences. Second, to show that we can modify traditional rules and hence devise new mathematics. And third, to show how we can produce new mathematics that serves to describe aspects of the physical world.

Each of these claims is true not only of arithmetic and algebra but also of other branches of mathematics, such as geometry and calculus. For example, nowadays many physicists believe that the structure of the universe does not quite correspond to the principles of ordinary ancient geometry. Indeed, anyone acquainted with questions of physical meaning in mathematics might immediately think of physical geometry as the main field of interest. Hence, many books and articles have been written about physical geometry. Likewise, many writers have discussed the physical significance of the calculus.

But here we will not discuss such matters, and instead will focus on elementary numerical and algebraic concepts. Just simple numbers and signs. Why? For three reasons: (1) Because basic important questions about the physical meaning of mathematical concepts are also present in the history and subject of numerical algebra. (2) Because such issues began to transpire in the development of the algebra of signed numbers before similar questions arose in calculus and modern geometry. (3) Because the subject of abstract numbers and elementary algebra is more easily accessible to general readers.

We will concentrate attention on negative numbers. Why? Because they are unassuming but fun. Simple paradoxical gems of the practical imagination. The long-neglected negatives stand as just about the only kind of numbers about which a book has not been written. And they suit the study of creative mathematics well because they lie precisely between the obviously meaningful and the physically meaningless. Thus we think about negative temperatures, but not about a negative width. By first studying interpretive disputes in the history of signed numbers, we will place ourselves on that borderline between the meaningful and the meaningless, and we will gradually pry open the old dusty crate of innovative mathematical representation.

Furthermore, whereas the subject of physical geometry has been investigated at length, by contrast, the analogous case for arithmetic and algebra remains obscure. Even nowadays. Do traditional quantitative methods serve to represent all physical magnitudes and relations exactly?

But wait, this is not a trick question. To answer it we will not turn to the abstruse or the technical. For example, philosophically minded readers might search in vain through these pages for terms such as intuitionism, idealism, physicalism, fictionalism, and so forth. Of course, any text can be given a philosophical reading, but the present book is not composed of philosophical subjects or opinions. Moreover, it does not discuss, nor even mention, not at all, seemingly intimidating physical concepts and subjects, like potentials, differentiable manifolds, isospin, negative energy, entanglement, gauge fields, phase space, or fiber bundles. Instead, we will deal with such very basic primitive concepts that it might seem as if we are scarcely even talking about physics at all. The examples discussed and the puzzles laid out will deal only with seemingly simple notions, such as quantity, direction, position, length, area, displacement, symmetry, and speed.

Accordingly, the present book deals mainly with the elementary treatment of numbers and variables as a simple way of introducing the study of custom-made mathematics. We will analyze old ambiguities in traditional rules that otherwise have remained neglected for a long time. We will develop new solutions to basic problems that were solved or dismissed long ago.

Now, of course, it is not a popular practice to work on questions that are not widely acknowledged as problems. Some people might shudder at the thought of seriously reviving discussions that are seldom encountered except in old books and history books. Certain sorts of superior persons believe that they were born *after* history. For example, they believe, for all practical purposes, that the elementary foundations of mathematics were discovered long ago, and established so securely that today we need not bother to question the validity of the simplest operations. They presume that if anything remained ambiguous or problematic in the elements of mathematics, then it would have been solved already by some clever fellow back in the nineteenth century.

But there are various reasons why problems cease to concern people. Oftentimes, yes, a problem is superseded because someone finds a solution that is generally satisfactory. Other times, however, a problem is not solved directly; people just find ways to circumvent it. And sometimes a particular problem ceases to be a focus of attention for more mundane reasons: individuals grow tired of the problem, their attention is diverted to other subjects, or they become satisfied with catchphrases and rhetoric that dismiss it as illusory or trivial. Once a problem has been discussed for a while, proposals for its solution or dismissal may be sufficiently attractive to convince newcomers that they need not pursue it in detail nor attempt to elucidate the matter any further. In the meantime, those who were originally attempting to find or refine a solution, or those who at least openly recognized the significance and persistence of the problem, eventually die. Without newcomers continuing to analyze the problem, the impression can emerge that the problem was solved or was merely apparent. But with any such vague impression, we may ask, exactly who solved the problem, when, where, and *how*? If such questions are met merely with a dismissive shrug of the shoulders or with the innuendo that there merely *seemed* to be a problem only to people of the past who were rather stupid or confused, then we may suspect that the problem actually remains unsolved. Perhaps people just learned to live with it. And even if we are told how the problem was allegedly solved oh-so-long-ago, we may nonetheless analyze the solution to see if it is really satisfactory *to us*.

It is well known that to advance science and mathematics researchers should carefully study contemporary and recent works in their field. By contrast, the study of the distant past rarely receives close attention, except from historians. But old works and ideas are not always irrelevant to contemporary discussions. We should not assume that all that was written long ago ever received its proper share of attention. We cannot assume that everything of value in old books has effectively been carried into the present in subsequent works. Any acquaintance with history can show that sometimes fertile ideas are forgotten or neglected. Depending on the subject of investigation, we may find that more analyses and commentaries were produced in the distant past rather than the recent past. Hopefully, the present study will demonstrate that studying the distant past can serve also for the advancement of scientific understanding.

So we will reconsider old questions and controversies buried in the history of mathematics. Happily, the discussions and historical excursions that follow focus on basic mathematical procedures familiar to most people. Otherwise, nonspecialists who begin reading books on mathematics or its history often lose interest quickly when such books abruptly discuss complicated subjects entirely alien to them. Furthermore, the historical portion of this book, unlike many works in the history of mathematics, is not a history of success. It is not an account of how individuals solved one perplexing problem after another. Instead, the historical excursions are meant to convey a sense of the ambiguities inherent in ordinary mathematical rules, ambiguities that have been made invisible by generations of textbook writers.

Finally, a word should be said about the style and format of this book. It could have been unreadable. But what for? Already plenty of seemingly incomprehensible books sit on library bookshelves, enough to last anyone a lifetime. Big words in small print, like thick brick walls separating the expert from the innocent; so many technical books. So let this one read easily instead. And if, along the way, mathematically expert readers complain (as they should) that the book lacks thoroughness because it doesn't trace out the intricate consequences of certain propositions, then they can well pursue such questions further. Mathematicians have the skills to read many texts that for other people are virtually inaccessible. They can make sense of this one as well. Meanwhile, it should be clear enough for anyone who knows at least some elementary mathematics.

chapter 2

The Problem

Despite the great usefulness of mathematics in science, mathematics involves rules that do not correspond to our ordinary experience.

This is not a problem for mathematics in general, because many fields of mathematics do not aim to describe relations among observable things. But it can become a problem, at least when we use math as a tool for describing various things exactly.

For example, the mathematics commonly used in physics involves a mixture of elements that vary widely in their physical significance, including some that seem meaningless. As it is, even elementary math involves rules that hardly relate to experience. Happily, however, we have some significant freedom to develop new math rules as we see fit.

We can improve the tools for representing and analyzing phenomena. This can be achieved directly or indirectly. The latter path is sometimes evident when some abstract mathematical systems find useful physical applications, though originally they were not conceived for such purposes. The math comes first, and then it sort of sits on a shelf waiting to be employed eventually. Yet we can actively pursue the former path as well: to develop concepts and methods that seek directly to describe particular structures of things and to serve specific needs.

For the purpose of this book, there are three main reasons for

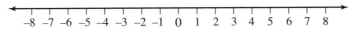

Figure 1. A number line.

tampering with certain rules. First, some such rules lead to peculiar asymmetries. Second, such rules lead to complications that can be avoided. And third, they entail some symbolic statements that are practically meaningless.

These reasons will now be illustrated. But to best understand them, the reader is now asked to momentarily consider the subject afresh, as if at first acquaintance with the rules in question. This exercise should highlight ambiguities that otherwise have been made invisible in introductory textbooks for a long time. There are many ways of introducing a subject, and of course educators want to magnify simplicity. But despite their good intentions, the resulting apparent clarity can be deceptive. Now, by focusing instead on ambiguities, the reader can develop empathy for past mathematicians as they struggled to make sense of the subject. The history of the controversies that for centuries plagued the use and interpretation of negative and imaginary numbers is reviewed in the next chapters. The following discussion illustrates peculiarities that are nowadays disregarded—though they persist, not only as the problems motivating this inquiry but also as reasons why many people have difficulties understanding math.

First, consider the claim that some ordinary rules lead to asymmetries. Consider a number line as shown in figure 1. This figure seems to suggest a perfect *symmetry* among negatives and positives as merely numbers to the left or right of zero. However, certain simple operations on such numbers do not give symmetric results. For example, consider the operation of squaring. The square of any number to the right of zero yields a number that is *also* to the right of zero:

$$(2)^2 = 4,$$

whereas the square of a number to the left of zero does *not* yield another number also to the left of zero:

$$(-2)^2 = 4.$$

If you are new to these matters, or if you happen to have forgotten the established rules of exponents, you may think "wait, maybe this asymmetry results because we are carrying out the *positive* square of the negative number." What happens when instead we consider negative squares? But again, we do not obtain negative results. Instead, the results may seem even more peculiar, as for example,

$$(-2)^{-2} = \frac{1}{4},$$

and likewise,

$$(2)^{-2} = \frac{1}{4}.$$

These are formal asymmetries; all these results seem to privilege positive numbers over negatives. What do these operations mean in terms of the number line? Moreover, stranger things happen with the extraction of roots. The square root of a number to the right of zero may also be a number to the right of zero,

$$\sqrt{4} = 2,$$

but the square root of a number to the left of zero does *not* yield a number to the left of zero,

$$\sqrt{-4} = 2\sqrt{-1},$$

or

$$\sqrt{-4} = 2i.$$

The number $2i$ is not negative. It is not positive either. This number, like all other "imaginary" numbers, is neither greater than zero, nor less than zero, nor even equal to zero. It is a number that is not located *anywhere* on the number line. Thus the asymmetry: the extraction of square roots from positive numbers yields results that are located on the number line, while the same operation with negative numbers does not. Other asymmetries, some of which will be discussed later, further illustrate the formal differences between positives and negatives, but these examples suffice for now.

So what? Who cares about symmetry, you may ask, old-fashioned architects? Actually, physicists care about symmetry because we experience it all around us. Walking to the right seems to be no more special than walking to the left. Does it? Yet math treats positives and negatives differently.

Consider now how the rules in question entail complications. For example, early writers of mathematical texts established that multiplying two negative numbers should result in a positive. Thus we say that $-2 \times -2 = 4$. Of course, the multiplication of two positive numbers was also deemed to result in a positive, for instance, $2 \times 2 = 4$. Hence, the extraction of the square root of any number entailed *two* results. For example, strictly speaking,*

$$\sqrt{4} = \pm 2.$$

Thus originated a rarity in arithmetic: a simple operation that yields not one but two solutions. By contrast, other operations, such as $2 + 2$, have a *unique* solution. Thus, the ambiguity inherent in double solutions constitutes a complication that, just maybe, might otherwise be avoided. If maybe it were possible to establish, say, that

$$-2 \times -2 = -4,$$

then the complication of multiple solutions would disappear since an expression such as $\sqrt{4}$ would accordingly have only *one* possible solution.

Another complication stems from the consequent question of what is the square root of a negative number. Since the squares of both negative and positive numbers all were deemed to be pos-

---

* Notice that here the square root symbol is distinct from the one used before, as it lacks the line or "vinculum" over the radicand. Owing to the ambiguity of double solutions in the square root operation, mathematicians convened to use the sign $\sqrt{}$ to indicate the operation of extracting only the so-called "principal" root. We follow that convention while using the sign $\sqrt{}$ to designate the unrestricted root operation. For other alternatives see Florian Cajori, *A History of Mathematical Notations*, vol. I (Chicago: Open Court, 1928), 378–379.

itive, mathematicians invented a new sort of number in order to express roots of negative numbers. Thus began the practice of writing

$$\sqrt{-4} = 2i.$$

Clearly, the solution of appealing to a new kind of number is a complication that might well be avoided if somehow instead we could establish simply that, say,

$$\sqrt{-4} = -2.$$

If so, then perhaps we would have no need for imaginary numbers. What happens if we actually experiment and make such irreverent changes in these basic rules of signs? That is a question that we will later investigate.

Finally, consider what is actually the major reason motivating this project of tinkering with traditional mathematical principles: that in some practical contexts some rules produce meaningless or ambiguous results. Traditionally, we are taught how to apply mathematical rules and concepts in specific ways that generate meaningful results. However, examples can also be constructed easily where instead the same rules and concepts yield results that are physically meaningless.

For example, imagine that Tom has five apples and he gives three to Amy; how many apples does Tom have left? If we say "two," we're all quite happy with the answer and the efficiency of mathematical subtraction. But what if Tom has five apples and he gives *eight* to Amy; how many apples does he have left? The answer "negative three" is unsatisfactory. Why? Because we know from experience that despite the happy notion of numbers less than zero, you cannot have less than zero apples. The operation of giving away eight apples cannot be completed if you only have five. The ambiguity can be disguised by inventing a new interpretation for the result, say, that the number −3 represents the quantity of apples that Tom *owes* Amy. But then we have changed the nature of the problem. Instead, just perhaps, we could change the rules underlying its mathematical formulation in order to express the physical fact that you cannot give away more than you have.

Nonetheless, someone might insist that such ambiguities are really just questions of interpretation, and that they disappear by learning how to interpret mathematical results "properly." Unfortunately, even if this is somehow argued for negative numbers, it is quite untenable for imaginary numbers. For example, Tom has $4i$ apples and he gives three to Amy; how many apples does he have left? Who knows! The problem has no solution. Here, the simple arithmetical operation of subtraction cannot be performed. The concept of having an imaginary quantity of things makes no physical sense. Consider another example: Augustus is very hungry—he hasn't eaten for days—and we say: "we can give you either two candy bars or $2i$ candy bars; which would you prefer?" How can anyone answer that? It is entirely unclear which is the greater quantity and which the smaller. We do know that 2 and $2i$ are not the same because $2\sqrt{-1}$ is not equal to 2. Yet $2i$ is also neither greater than zero nor less than zero. So how do we assess Augustus' options? We can't elucidate the decision because the imaginary offer is meaningless. The simple quantitative comparison between the two numbers cannot be carried out.

Another way in which the elementary mathematical concepts do not correspond to physical relations concerns the asymmetries between positives and negatives. For example, suppose that we use positive numbers to represent motions to the right along a line, and negative numbers to represent motions to the left. Physically, there is nothing special about the "positive" direction by comparison to the "negative." But the traditional rules on the operation of positive and negative numbers involve peculiarities that do not correspond to the physical symmetry. For example, the cube root of a motion of eight steps to the right is positive:

$$\sqrt[3]{+8} = +2,$$

just as the cube root of a motion of eight steps to the left is negative:

$$\sqrt[3]{-8} = -2.$$

Likewise, the square root of four steps to the right can be positive:

$$\sqrt[2]{+4} = +4.$$

However, the square root of four steps to the left is not negative:

$$\sqrt[2]{-4} = 2i.$$

Owing to the meaninglessness of such results, physicists often and routinely disregard imaginary roots of equations, as well as the negative roots that can be obtained from positive numbers. Other applications of the elementary rules also entail meaningless asymmetries. For example, we often use positive numbers to represent money that is possessed, and negative numbers to represent debts. But then, whereas the square root of a monetary asset can be said to be an asset too, the square root of a debt is not a debt; it is imaginary.

Furthermore, negative numbers also involve profound ambiguities in our understanding of numerical ratios. We are all comfortable with the proposition that

$$\frac{1}{2} = \frac{3}{6},$$

where both pairs of quantities have the same ratio. We also understand and accept the inequality

$$\frac{1}{2} \neq \frac{2}{1}.$$

The ratio of the smaller number to the greater is not equal to the ratio of the greater to the smaller. But consider the following proposition of ordinary mathematics:

$$\frac{-1}{1} = \frac{1}{-1}.$$

How can the ratio of a smaller quantity to a larger be the same as the ratio of the larger quantity to the smaller? Or differently put, how can one possibly divide a smaller number by a greater and get the same result as by dividing the greater number by the smaller? What does such a proposition mean physically, say, in terms of apples? *Really*, what does it mean? Moreover, the meaning of division involving imaginary numbers is even more ambiguous, but we will leave it at that. More examples of physically ambiguous mathematical propositions will be given later, throughout, but these few suffice for now.

Having now a sense of some of the peculiarities inherent in the elements of mathematics, we are in a position to empathize not only with students who have difficulties when they first encounter these subjects, but also with mathematicians, scientists, and philosophers who for centuries had similar concerns. Such concerns led to the development of new mathematical methods. One field that called for improvement was the mathematical methods of representation used in physics. Yet often the advancement of mathematics was carried out irrespective of physical significance. A history of the interpretive ambiguities that accompanied that advancement is the subject of the following chapters.

chapter 3

History: Much Ado about Less than Nothing

The general acceptance of negative numbers and their rules of operation was an exceedingly slow process. In Europe, negative numbers have been used widely at least since the mid-1500s, yet their use and interpretation involved much disagreement among mathematicians well into the middle of the 1800s. So it took roughly three hundred years; an anomaly in the history of mathematics, such a very long of time before the general acceptance of such apparently simple elements. Why was there so much resistance and controversy in connection with negative and imaginary numbers?

The common tendency in histories of mathematics is to downplay or underestimate the degree to which mathematicians debated ambiguities relating to negative numbers. Perhaps writers imagine that since negatives are so elementary in algebra their introduction should be treated with appropriate simplicity. But against the view that the principles of a discipline are its easiest parts, there is opposed the view that sometimes the establishment of those principles is most difficult, and afterward much else follows readily. Whatever the case, the following account does not jump from problems to their solutions but instead dwells on the continuity with which various individuals raised objections to traditional theory.

Unlike many popular accounts of episodes in the history of mathematics, the following is not a story of success. It is not an attempt to reconstruct the process by which mathematicians ascertained right solutions and reached consensus. Also, by no means is it an exhaustive or comprehensive account of all opinions and developments on a particular subject. Rather, it consists of a selection of some relevant points. What follows is an overview of the complexion of disputes in the advancement of the theory of negative quantities. It is meant to undermine any illusion that mathematicians easily reached consensus on the matter. The reader is encouraged to attempt to empathize with the conceptual difficulties that troubled past writers. Only then can we begin to understand why ideas that to us may seem so eminently simple actually led able and prominent mathematicians to radical disagreements. Hence, the focus is not on the problems and their solutions, though some will be discussed later, but on conceptual problems as manifested in mathematicians' expressions.

Throughout the following account, several themes emerge: questions as to the meaning of symbolical rules and expressions, the interrelations between experience, diagrams, and symbols (especially in regard to the representation of physical quantities and qualities), the development of mathematics as a "language" or a "science," the creation of new mathematics from dissatisfactions with prevalent concepts, and, the practical utility of various algebras. These themes should give some historical perspective with which to better understand the questions treated in the chapters that follow. Again, the history of ambiguities in the theory of signed numbers is pursued here essentially as but one example to show that the relationship between experience and mathematics has rarely been as univocal or straightforward as one might otherwise assume.

For centuries, many mathematicians, philosophers, and scientists raised various objections repeatedly concerning the use of negative numbers. Negative numbers were useful especially in commercial transactions, as a means for representing debts, and hence they acquired wide currency. However, mathematicians, such as

Nicholas Chuquet and Michel Stifel, often characterized negative numbers as "absurd."[1] Even those who helped to advance the general use of such numbers were uncertain about their validity and significance.

Girolamo Cardano, for example, became widely known for his *Great Art* of 1545, wherein he used the square root of "fictitious" or "negative" numbers to convey the impossibility of solving certain equations.[2] He described such roots as "sophistic negatives" and characterized them as "useless." But such expressions, as he demonstrated, could be useful at least in the study of algebra, irrespective of their impossibility in geometry and arithmetic. Other mathematicians followed his example. Yet in a lesser known work published subsequently, Cardano tried to reject such expressions altogether.[3] Instead of assuming, as usual, that multiplication of two "alien" quantities, that is, negative numbers, produces a positive, Cardano decided to establish that it produces a negative.[4] Hence, the square root of a negative number would be negative, and there would be no need to invent and employ sophistic roots. He also modified the rules of division accordingly. For Cardano, negative numbers were admissible at least because they could be used to represent real quantities such as debts.

Uncertainty about the meaning and validity of expressions involving quantities less than zero was one factor that led mathematicians, scientists, and philosophers to view algebra with suspicion and distrust. Meanwhile, geometry presided, as it had for centuries, as the paragon of mathematical clarity and truth. No lines had negative length. Still, as mathematicians translated propositions of ancient geometry into algebra, they mixed them with the relatively novel rules of signs. For example, to reduce a square having sides of length a to a smaller square having sides of length $a - b$, algebraists intermingled the rule that minus times minus gives plus with propositions set forth in Euclid's famous *Elementa*, a work written some 300 years before the Christian era. Some geometers objected to this kind of reasoning. For example, in the 1570s, Federico Commandino, an authoritative translator and commentator of ancient Greek mathematics, criticized those who used the rule that minus times minus is plus, rejecting it as "untrue," following Cardano.[5]

Some mathematicians avoided the ambiguities of negative numbers by restricting arithmetic and algebra to the solution of problems that give positive results. François Viète, for example, defined subtraction as the operation of taking a smaller quantity from a greater quantity.[6] But this approach was troublesome for two main reasons. First, Viète and others aspired to a symbolic art that would allow the solution of *every* problem, in which case expressions involving subtraction of a greater number from a lesser would require solution. Second, in algebra often one does not know which of two symbols designates the larger quantity, so that it is sometimes necessary to proceed in ways that involve subtraction of a greater from a lesser.

Algebra was generally understood as the study of the relations among quantities. Mathematics was sometimes defined synonymously as the "science of quantity" and as the "science of magnitude," at least until the mid-1800s. Hence, negative numbers were defined as "quantities less than nothing."[7]

In the 1600s, mathematicians continued to entertain various objections against negative numbers. Thomas Harriot usually avoided and dismissed equations involving the square roots of negative numbers as inexplicable and impossible. In manuscripts, he also modified the rules of multiplication of signed numbers, establishing, in particular, that minus times minus gives minus, to ascertain ways to solve equations without using impossible roots.[8] And, like Cardano, he rejected the negative roots of positive numbers. Blaise Pascal claimed that the result of any subtraction could never be less than zero. Antoine Arnauld questioned the significance of negative numbers, arguing that the equation

$$-1 : 1 = 1 : -1$$

was an impossible relation, because it indicated that the proportion of a smaller number to a greater was the same as the proportion of the greater number to the smaller.[9] Ordinarily, it would be ridiculous to assert that, say,

$$2 : 4 = 4 : 2.$$

This peculiarity in the theory of proportions puzzled other mathematicians too, as we will later see.

Like others, Pierre de Fermat and René Descartes did not trust negative numbers as solutions of problems. They each independently devised a way of applying algebra to the analysis of geometrical problems that became known as analytic geometry or coordinate algebra. But they generally avoided using negative numbers in their analysis of curves; in fact, they rarely plotted figures using negative coordinate points.[10] In his *Géométrie* of 1637, Descartes called negative roots of equations "false" (though he could convert them into "real" positive solutions), and he entirely rejected roots of negative numbers and so coined the term "imaginary."[11] Many of the early successors of Descartes and Fermat "continued throughout the century to make mistakes through overlooking or misinterpreting negative coordinates."[12] One exception was John Wallis, who usefully interpreted negative numbers as lengths opposite in direction to lengths represented by positive numbers. But Wallis also had objections against the notion of negative "quantities." He argued that "it is also Impossible, that any Quantity . . . can be *Negative*. Since that it is not possible that any *Magnitude* can be *Less than Nothing*, or any *Number Fewer than None*."[13]

To make sense of negative expressions, Wallis interpreted them not only as less than zero but also as greater than infinity.[14] To explain, consider a series of ratios: $\frac{24}{4} = 6, \frac{24}{3} = 8, \frac{24}{2} = 12$, etc. Thus far, each value, 6, 8, 12 is smaller than infinity, of course. But, as the denominators get smaller (4, 3, 2, . . .), the value of the ratio grows larger and larger. Continuing the progression, some mathematicians thought that if the denominator is 0 the ratio must be infinity. Therefore, Wallis expected that if the denominator is even smaller, *less than* 0, then the ratio would be *greater* than infinity. And, since it was established that, say, $\frac{24}{-1} = -24$, then negative values could seem to be greater than infinity. By the way, nowadays mathematicians and writers of textbooks avoid such conundrums by saying that division by zero is undefined, that it cannot be done. (Still, it's interesting to note that some calculators nowadays give "infinity" if you use them to divide a number by zero.) Anyhow, another way to make the same argument is to consider a series of ratios where the denominator gets increasingly larger.

The value of each ratio is thus ever smaller. To obtain a ratio having the value of 0, we expect the denominator to be infinite. Therefore, to obtain a value less than 0 we might expect the denominator to be greater than infinity. Thus it seemed that passage to the negative realm could be through zero or through infinity. Subtraction from 0, as well as division by numbers greater than infinity, seemed to produce negative numbers.

As for "the *Supposed* Root of a Negative Square," Wallis proposed several ways to make sense of the concept geometrically.[15] For example, he showed how to use it to stand for the side of a negative square area, such as the side of a square acre lost to the sea. Also, he showed that an imaginary expression could be interpreted as a point outside a line containing positives and negatives, a tangent line, a point outside a plane, and, furthermore, to denote extensions to problems, such that if a problem has no solution for a circle it may apply, say, to an ellipse. Thus, he showed that imaginaries could be ascribed geometrical significance. Nonetheless, he admitted that, first of all, in the strict original formulation of a problem, such expressions served to signify impossibility. Likewise, other mathematicians, among them Isaac Newton, designated the square roots of negative quantities as "impossible."[16]

Mathematicians cautioned that abstract concepts should be applied to physical considerations carefully. Nicholas Saunderson, in his *Elements of Algebra* of 1740, noted that negative quantities could be used meaningfully in some physical contexts, such as in discussions of monetary debts, temperatures, and motion. Yet Saunderson also admitted that "The possibility of any quantitie's being less than nothing is to some a very great paradox, if not a downright absurdity; and truly so it would be, if we should suppose it possible for a body or substance to be less than nothing."[17] Likewise, to explain the rules of negative quantities, Colin MacLaurin relied mainly on the notion of contrariety.[18] Given two quantities in some way contrary to one another, such as motions or forces in opposite directions, MacLaurin explained that their combination could be described by the rules of addition and subtraction of positive and negative numbers. Thus the "quality," positive or negative, resulting from an addition of two quantities, would be the same as the quality of the larger of the two. But MacLaurin noted that this mathematical rule did not serve for the analysis of just any physical quantity:

> This Change of Quality however only takes place where the Quantity is of such a Nature as to admit of such a Contrariety or Opposition. We know nothing analogous to it in Quantity abstractly considered; and cannot subtract a greater Quantity of Matter from a lesser, or a greater Quantity of Light from a lesser. And the Application of this Doctrine to any Art or Science is to be derived from the known Principles of the Science.[19]

In algebra, one might say that negatives were "less than nothing," but this notion and its associated operations would be valid only in *some* physical applications, not generally.

MacLaurin justified the introduction of signed numbers into algebra on the basis of physical utility. He argued that mathematics dealt not only with concepts of magnitude but also with other concepts having physical significance. Thus, signed numbers served to represent and analyze positions, motions, directions of forces, brightness of optical images, gravity, and more. Hence he considered negative quantities as "no less Real" than positives.[20]

Since expressions involving the square roots of negative numbers did not correspond to solutions of real problems, MacLaurin characterized such expressions as impossible. He argued that the expression $\sqrt{-1}$ did not represent a quantity or number but an impossible operation supposed to be carried out on a real quantity.[21] Still, he noted that in some cases the combination of imaginary expressions could serve to compensate, that is, eliminate the impossibility.[22] For example, the sum of $1 + \sqrt{-1}$ and $1 - \sqrt{-1}$ constitutes a real result, 2, because the impossibilities compensate one another. Still, he acknowledged that how such a compensation occurs is not obvious in every case. At least imaginary expressions served as useful though indirect ways to obtain real solutions for problems.

By the mid-1700s, the idea that reason could successfully serve to illuminate even great mysteries, and thus eliminate nonsense and mysticism, was growing into a prominent movement. Yet the notion of quantities or magnitudes "less than nothing" still seemed quite "absurd" to some thinkers of the Enlightenment.[23] The mathematician and physicist Jean-Le-Rond d'Alembert believed that the concept of negative numbers was one of the major fundamental problems of mathematics, along with the ambiguous status of the concept of parallel lines in geometry, and the notion of infinity in the calculus. In the famous *Encyclopédie*, d'Alembert argued that a quantity less than zero is inconceivable, and that it should be rejected because −1 is to 1 as 1 is to −1; the two have the same relation.[24] The statements

$$-1 : 1 :: 1 : -1$$

and

$$\frac{-1}{1} = \frac{1}{-1}$$

would not be true if −1 were less than zero.[25] We can understand his claim by interpreting the quantities as the same while the qualities are different, for instance, as one apple is to one orange as one orange is to one apple.

Mistakes were widespread in the use of negative numbers if only because there was hardly any consensus on their significance. One way to interpret such ambiguities was to assume that negative numbers corresponded to mathematical impossibilities. Thus, in the *Encyclopédie* and elsewhere, d'Alembert argued that a negative solution "indicates that a false supposition was made in the formulation of the problem."[26] Thus the equation

$$100 + x = 50,$$

where the solution is $x = -50$, really meant, he claimed, that the problem had been stated incorrectly. Instead, the "false supposition" that a number *added* to 100 equals 50 should be restated to say

$$100 - x = 50,$$

such that $x = 50$, and so the negative solution is unnecessary.[27] Moreover, d'Alembert emphasized that there is really no such thing as an isolated negative quantity, such as -3, that 3 is the quantity properly speaking, though such an expression could be interpreted to mean, say, that someone *owes* three coins.

Though most of the rules on the operation of negative numbers were widely accepted, there were many disagreements over the justification of such rules. For instance, to justify the rule that $-a \times -b = +ab$, d'Alembert argued in the *Encyclopédie* that there were tacit errors in the formulation of the problem, such that $-a$ and $-b$ should have been $+a$ and $+b$, so that $+a \times +b = +ab$. Note, however, that this explanation might be unsatisfactory since it clashes with the other rules of multiplication: it would suggest that $+a \times -b = +a \times +b = +ab$, instead of $+a \times -b = -ab$.

As another example, consider the question of the justification of the rules of division with negatives. In a memoir titled "On the negative quantities," d'Alembert explained in passing: "if someone demands why $\frac{aa}{-a} = -a$, I will reply that by dividing the quotient of the division of aa by $-a$, one does not ask how many times $-a$ is contained in aa, that which would be absurd, one asks a quantity such that being multiplied by $-a$ gives aa."[28] But for centuries division had been conceived as concerning the very question of how many times one quantity fits into another. Thus, because a traditional explanation of division led to absurd notions when applied to negative numbers, it was summarily replaced with another that hid the ambiguity by reformulating division as multiplication.

Interpreting the results of algebraic problems was even more difficult when the square roots of negative numbers were involved. The symbol $\sqrt{-1}$ had no physical meaning, ambiguous geometrical meaning, and no arithmetical meaning. Thus some theorists sought to explain it by means of metaphysics. After all, impossibility, "all that involves contradiction," was a question deemed to pertain properly to metaphysics.[29] But, for practical purposes, imaginary solutions were routinely discarded.

In the mid-1700s, at the city of Danzig, in Prussia, a teacher of mathematics, Heinrich Kuhn, proposed a way to construe the square roots of negative numbers in terms of geometrical constructions.[30] Kuhn argued that just as the expressions $+a$ and $-a$

may be taken to designate opposite segments of a horizontal line bisected by a perpendicular line, so also the two segments of the perpendicular line may be designated by $+a$ and $-a$, if only the lines are of equal length. He then claimed that the four squares determined by the perpendicular lines could be designated by expressions conveying the roots of each square area. He designated any side of the upper right square by the expression $\sqrt{+a^2}$. Accordingly, he designated the sides of the bottom right square by $+\sqrt{-a^2}$. He designated the sides of the bottom left square by $\sqrt{[-a \cdot -a]}$ or $-a$. And he designated the sides of the upper left square by $-\sqrt{-a^2}$. Thus, Kuhn illustrated a deliberate way of associating geometrical figures with negative and imaginary expressions, and he proceeded to analyze several square and rectangular figures on that basis.

Like Kuhn, many mathematicians believed that, although negative and imaginary solutions were often difficult to understand, the ambiguities would be resolved by finding proper ways to interpret such solutions. A few, by contrast, pursued a more radical approach: they concluded that the ambiguities would only be resolved, really, by discarding some part of the procedures that led to them. The latter path was taken by Francis Maseres in his *Dissertation on the Use of the Negative Sign in Algebra* of 1758.

Like other mathematicians, Maseres realized that algebra lacked the certainty of geometry. He conjectured that the clarity and accuracy so widely admired in geometry had been "almost universally neglected in books of Algebra," partly because of

the greater facility with which a writer may impose upon himself as well as his readers, and fancy he has a meaning where, in reality, he has none, in treating of an abstract science, such as Algebra, that addresses itself only to the understanding, than in treating of such a Science as Geometry, that addresses itself to the senses as well as to the understanding; for the impossibility, or difficulty, of representing a false, or obscure, conception in lines to the eye, would immediately strike either the writer or the reader, and make them perceive its falsehood or ambiguity; whereas when things are expressed only in words, or in any abstract notation, wherein the senses are not concerned, men are much more easily deceived: and for same reasons, a defect of proof, or a hasty extension of a conclusion justly drawn in one case to several other cases that bear some resemblance, but not a complete one, to it, may be more easily perceived in Geometry than in Algebra.[31]

Accordingly, Maseres sought to introduce into algebra a degree of rigor and exactitude comparable to that of geometry. Since antiquity, geometrical knowledge, especially as formulated in the *Elementa* of Euclid, had been admired greatly for the apparent truth and rigor with which propositions were derived from one another. Following this model, Maseres sought to lay out the steps of algebraic reasoning explicitly, in order to attain demonstrative clarity.

Maseres attributed the ambiguities and lack of rigor in algebra mainly to the careless introduction and use of the concept of negative quantity. He argued that the use of the plus and minus signs in connection to any single quantity instead of to a pair of quantities "must be mere nonsense and unintelligible jargon."[32] For example, he demonstrated the rule "$- \times -$ gives $+$" as having meaning only when each minus sign is shared by pairs of quantities, rather than attached to an isolated quantity. Thus, the multiplication of single signed numbers, such as -3×-4, would be meaningless, whereas $(5 - 3) \times (5 - 4)$ would be meaningful. Accordingly, he advocated the use of the minus sign to denote only "the subtraction of a lesser quantity from a greater."[33]

In particular, Maseres rejected negative roots of equations. For example, instead of accepting the common idea that a quadratic equation, such as

$$x^2 + 2x = 15,$$

has two solutions (in this case $x = 3$ and $x = -5$), he argued that this assertion really concerned two distinct equations:

$$x^2 + 2x = 15 \quad \text{and} \quad x^2 - 2x = 15,$$

each having one solution (in this case, 3 and 5, respectively). Thus he demonstrated how to solve different quadratic equations independently, rather than generally. He likewise demonstrated this same sort of individualized analysis for cubic equations. Instead of allowing that multiple equations be united into one equation having multiple solutions, he insisted on treating each distinct case as a single equation having one solution. Otherwise, he commented, multiple roots "serve only, as far as I am able to judge, to puzzle the whole doctrine of equations, and to render obscure and mysterious things that are in their own nature exceeding plain and simple." Accordingly, he mused:

> It were to be wished therefore that negative roots had never been admitted into algebra, or were again discarded from it: for if this were done, there is good reason to imagine, that the objections which many learned and ingenious men now make to algebraic computations, as being obscure and perplexed with almost unintelligible notions, would thereby be removed; it being certain that algebra, or universal arithmetic, is, in its own nature, a science no less simple, clear, and capable of demonstration, than geometry.[34]

By rejecting negative numbers, Maseres also rejected imaginary numbers.

Though presumably few other mathematicians sought to solve the ambiguities of algebra by discarding negative numbers entirely, at least many felt it necessary to elucidate the significance of negatives and imaginaries. For example, François Daviet de Foncenex noted that: "So often one finds imaginary quantities in algebraic expressions that it makes one wish that we were dedicated to examining with more care their nature & origin. Such researches would be of great help in all parts of Mathematics treated by calculation, & thus one would have avoided many paradoxes & contradictions in a Science that should be entirely exempt of such."[35] Like other writers, Daviet de Foncenex noted that the presence of imaginary roots in the solution of problems implied that contradictions had been committed in the formulation of such problems. In view of such contradictions, he concluded that, "evidently," geometrical constructions could not be made from such problems. He rejected as useless the idea of representing imaginary solutions by means of a line perpendicular to another line representing positives and negatives.[36] Nonetheless, he argued that whereas some imaginary expressions constitute impossible or "foreign" cases of a given question, in other cases an imaginary expression was actually a roundabout alternative way of expressing a meaningful formula. In his words, "often, a formula that apparently should be entirely satisfactory to a question, we present it instead in certain circumstances in an absurd and impossible fashion."[37] Thus he sought to identify cases in which apparently impossible or contradictory expressions were but artifacts of particular algebraic formulations.

Meanwhile, in German-speaking lands, negative numbers and impossible expressions were increasingly gaining acceptance. For example, in 1763 the philosopher Immanuel Kant published his "Attempt to Introduce the Concept of Negative Quantities into World-Wisdom." Kant deemed that it was useless to attack the mathematical theory of negatives by pitting metaphysical principles against it. Instead, he proposed to inquire whether the mathematics of negative quantities could serve to help bring order to the tangles of philosophy.

Kant distinguished between two kinds of oppositions. Logically opposite things were absolutely incompatible, whereas things that are opposite in reality can and do actually coexist. As an example of the latter, he considered two equivalent forces pushing a body in opposite directions; their result is real and presents itself as rest. He also explained that the signs + and − did not always mean addition and subtraction, respectively. An expression such as

$$-4 - 5 = -9$$

signified "no subtraction at all, but instead, a real augmentation and joining of magnitudes of one kind."[38] A quantity bearing the sign + would mean addition only when accompanying another quantity bearing also the + sign. Otherwise, different signs would just cancel one another insofar as their quantities are equal. The words of mathematicians, therefore, were not exactly adequate. Many "negatives" were not quite denials but, instead, oppositions between real things. Kant discussed various subjects: debts as negative capital, pleasure and unpleasure, polarities of heat and cold, the world and the divine will, and overall, the importance of distinguishing between logical and real oppositions.

Leonhard Euler, in his *Complete Introduction to Algebra* of 1770 (widely esteemed as the best algebra textbook for years), illustrated the significance of positive and negative quantities by appeal to notions of monetary assets and debts. Euler claimed that, since debts can be represented by negative numbers, one may therefore say that negatives are "less than nothing." Accordingly, he justified the rule that a negative number multiplied by a positive produces a negative, on the grounds that a debt taken several times still produces a debt. Although he apparently realized that his account was rather cursory, he acknowledged that: "It is of the utmost importance through the whole of Algebra, that a precise idea should be formed of those negative quantities, about which we have been speaking."

As for imaginary numbers, Euler described them as follows:

> The square roots of negative numbers, therefore, are neither greater nor less than nothing; yet we cannot say, that they are 0; for 0 multiplied by 0 produces 0 and consequently does not give a negative number.
>
> And, since all numbers which it is possible to conceive are either greater or less than 0, or are 0 itself, it is evident that we cannot rank the square root of a negative number amongst possible numbers, and we must therefore say that it is an impossible quantity. In this manner we are led to the idea of numbers, which from their nature are impossible; and therefore they are usually called *imaginary quantities*, because they exist merely in the imagination.[39]

Euler claimed that imaginary quantities were "of the greatest importance" for the analysis of questions where one does not immediately know if they involve any real or possible solutions. Thus, although he employed imaginary numbers, he distinguished them from numbers that seemed to exist in some real sense. Most other mathematicians shared this feeling, and thus when they characterized such strange numbers as "impossible" or "imaginary," these were not merely idle words, they *meant* them.

In addition to the interpretive ambiguities, there were also disagreements about the actual results of some operations. The major disagreements concerned logarithms. We will not discuss these developments in any depth, since they are beyond the scope of elementary arithmetic and algebra, but we must at least mention them since they engendered much confusion and debate.[40] Remember that a logarithm is simply the exponent to which a number is raised to become another; for example, since $3^2 = 9$, then "2 is the logarithm of 9 to base 3": $\log_3 9 = 2$. Accordingly, the following equations:

$$(+3)^2 = +9$$

$$(-3)^2 = +9$$

$$(+3)^{-2} = +\frac{1}{9}$$

$$(-3)^{-2} = +\frac{1}{9},$$

where each of the results is positive, can each be reformulated to show how to extract logarithms of *positive* numbers. But what about the logarithms of negative numbers? Thus mathematicians disagreed over the solution of problems such as

$$\log_3 -9 = ?$$

Gottfried W. Leibniz, for example, argued that negative quantities have no logarithms. Johann Bernoulli argued that the logarithms of negative numbers were the same as the logarithms of positives:

$$\log_b(-a) = \log_b(a).$$

Euler, however, subsequently established that for any positive number a, the logarithm of the corresponding negative is

$$\log_b(-a) = \log_b(a) + \sqrt{-1}(2k-1)\pi,$$

where k can be any positive whole number. Since k can have infinitely many values, then Euler had established that negative numbers have an infinity of logarithms, all with imaginary values.

Euler also established that positive numbers, too, have infinitely many imaginary logarithms, *but also* one positive logarithm. D'Alembert was not convinced by Euler's approach, suggesting that although it was an algebraically consistent solution it involved a degree of arbitrariness; that one might instead opt for a simpler and more natural solution. For example, he argued that one could establish that $\log_2 -1 = 0$. Daviet de Foncenex essentially accepted Euler's solution, but argued that negative numbers also had real solutions; sympathizing with some of the arguments of Bernoulli and d'Alembert. In turn, his arguments were criticized by d'Alembert. And so on. In the long run, Euler's approach was widely accepted as *the* solution, although debates continued, as mathematicians tried to ascertain also the logarithms of imaginary expressions.

In a paper presented to the Royal Society of London in 1778, John Playfair reviewed the perplexing status of imaginary numbers and attributed the confusions in algebra to "the doctrine of negative quantity and its consequences." Playfair contrasted the obscurities, controversies, and metaphysical disputes in algebra to the clarity of geometrical knowledge:

> The paradoxes which have been introduced into algebra, and remain unknown in geometry, point out a very remarkable difference in the nature of those sciences. The propositions of geometry have never given rise to controversy, nor needed the support of metaphysical discussion. In algebra, on the other hand, the doctrine of negative quantities and its consequences have often perplexed the analyst, and involved him in the most intricate disputations.[41]

Like other writers, Playfair attributed the origin of imaginary expressions to contradictions taking place in the combination of mathematical ideas. He noted that "The natural office of imaginary expressions is, therefore, to point out when the conditions, from which a general formula is derived, become inconsistent with each other. . . ."[42]

He acknowledged, however, that in some cases imaginary expressions could also be used for other purposes, such as to denote real quantities indirectly. In particular, geometers had found ways to represent circular arcs as well as hyperbolic areas by means of imaginaries. For example, for a circle of radius r, Bernoulli had represented its circumference by

$$2\pi r = \frac{4\log_2 \sqrt{-1}}{\sqrt{-1}} r.$$

Playfair commented that by employing such expressions one could derive "just conclusions," not by reasoning with ideas but by the mechanical manipulation of characters. Continuing to contrast algebra to geometry, he argued:

> But though geometry rejects this method of investigation, it admits, on many occasions, the conclusions derived from it, and has confirmed them by the most rigorous demonstration. Here then is a paradox which remains to be explained. If the operations of this imaginary arithmetic are unintelligible, why are they not altogether useless? Is investigation an art so mechanical, that it may be conducted by certain manual operations? or is truth so easily discovered, that intelligence is not necessary to give success to our researches?[43]

To solve the paradox, Playfair rejected the claims of Bernoulli and Maclaurin: that when imaginary expressions serve to facilitate useful results in calculation it is because they compensate or destroy one another. Playfair complained: "how can we conceive one impossibility removing or destroying another?"[44] Instead, he sought to ascertain the circumstances in which imaginary expressions lead to meaningful results.

Arguing that imaginary expressions only had geometrical meaning when describing properties shared by measures of angles and of ratios, Playfair proposed a "principle of analogy" to trace the affinity between such measures. He demonstrated that certain propositions valid for hyperbolic sectors were also valid for circular arcs. Thus he advocated the arithmetic of impossible quantities as a method for tracing analogies between circles and hyperbolas. Although there remained instances in which the analogy was not valid, at least some imaginary expressions admitted geometric interpretation.

While some mathematicians attempted to clarify algebra on the basis of geometry, others argued that algebra should be independent of geometry. Despite the success of Descartes' mixture of geometry and algebra, it was subsequently developed with greater focus on algebraic expressions than on diagrams of figures. In France several prominent mathematicians sought to emancipate algebra fully from geometrical constructions. Gaspard Monge labored to elucidate a system of coordinate geometry that was purely arithmetical. Joseph-Louis Lagrange devoted much work to replacing geometric figures with analytic formulas, especially in the field of solid geometry, and suggested that geometers build a completely analytic form of geometry. His student, Sylvestre François Lacroix, carried out this program, arguing that algebra and geometry "should be treated separately, as far apart as they can be. . . ."[45]

As mathematicians and natural philosophers relied increasingly on algebraic formulas without reference to geometric diagrams, the study of physics became increasingly abstract. With the proliferation of algebraic representations, theorists became increasingly convinced that all ideas of quality could be reduced to ideas of quantity, and hence that "every phenomenon is logically susceptible of being represented by an *equation*."[46] This conviction grew to the extent that geometry came to be viewed as superfluous to the exact analysis of nature. Hence, in 1788, Lagrange published a purely analytic formulation of mechanics, titled *Analytical Mechanics*, in which he boasted that he had developed the subject completely without appeal to a single diagram. He wrote: "One will not find any figures at all in this work. The methods that I present here require neither constructions nor geometrical or mechanical reasonings, but solely algebraic operations, carried out in a regular and uniform march."[47] Lagrange's work served as a demonstration that algebra and calculus sufficed for the analysis of physical problems without appeal to traditional geometrical methods. However, much of algebra had not been invented originally for the solution of physical problems. So it was not a fore-gone conclusion that algebra would be a precisely suitable tool for physics. Some of the principles of algebra were hardly understood in physical terms. Algebra thus appeared as a highly effective system devoid of logical foundations, like "a full-grown tree with many branches but no roots."[48]

THE SEARCH FOR EVIDENT MEANING

We now interrupt the historical account to highlight key issues relevant to our goals. We have begun to look at the history of conceptual difficulties in the use of negative numbers and imaginary expressions—but why? Why focus on "impossible" numbers and ambiguities of meaning in certain algebraic expressions? Why not instead illustrate any of the countless examples that show the utility of mathematical concepts and methods in science?

Well, first of all, you can easily find a wealth of such examples in any physics textbook and in many other sources. What most books do not exhibit, however, are the areas where mathematical concepts do not quite correspond to the physical things and relations we seek to describe. For the most part, teachers are quite busy teaching some of the countless useful applications of math for solving physical problems, while a complementary labor remains neglected: that of explaining to students the limits of applicability of specific mathematical concepts. Much time is invested studying applications where preestablished concepts are extremely useful, whereas hardly any time is invested in charting the cases or conditions in which experience and mathematics fail to correspond neatly to one another. Yet the latter task is also important.

We must realize that, despite the high usefulness of mathematics in analyzing phenomena, not all physical relations can be described by the same basic concepts. For example, negative numbers serve to convey perfectly meaningful physical relations in some cases, but in some others they do not: ". . . less of spent is more of saved; to lighten one's burden is to add to one's strength;

taking away cold is making warm. But there are problems where negative numbers are nonsense. A man cannot live a negative number of years. A pond cannot be −4 feet deep. A table cannot have −3 legs. No one is −6 feet tall."[49]

With the advance of science, increasingly many more uses have been found for negative numbers. So if at first the notion of debts was one of the few ways to make the concept useful, now we have no shortage of useful applications. Yet by studying the concerns of past scientists and mathematicians we are reminded that there yet remain many cases in which such numbers are meaningless. Accordingly, we may wonder why we continue to use the same concepts indiscriminately in situations where they are not quite suitable. In turn, we may attempt to develop new mathematical methods of representation and analysis.

So consider the expression −1. Expressions such as this one were used by merchants and accountants in various cultures to designate debts. Yet many early mathematicians did not accept such expressions as representing anything truly mathematical, like a number, because the concept of "a quantity less than nothing" seemed ridiculous. Nonetheless, since such expressions were very useful in commercial transactions and accounting, more and more writers of mathematical treatises began to include them. Thus, writers such as Cardano and Euler justified the use of negative numbers owing to the practical concept of debts.

But algebra was used not only when treating problems about money and assets, it was also used in many other contexts. Hence, once negative numbers were generally admitted, mathematicians and philosophers tried to find additional ways to make sense of these expressions in other practical contexts. A major difficulty, again, was that spatial magnitudes, such as width or height, could not be negative. But in several other contexts, especially those in which physical quantities admit of some kind of opposition, negative numbers found meaning. For example, Saunderson and MacLaurin explained that negative numbers were useful when discussing subjects such as motion, temperature, and more, but they cautioned that the algebraic rules on negative numbers should be used carefully, lest one apply them to contexts where they would be inappropriate and meaningless. In particular, the size of a material body, or a quantity of light, could not possibly be negative. MacLaurin cautioned that the mathematical theory should only be applied in each particular science *in accord with the known principles of that science.*

The case of MacLaurin is especially instructive because he justified the introduction and use of negative numbers precisely because of their physical utility. Traditionally, mathematics dealt with magnitudes, expressed by lines or numbers, yet MacLaurin argued that mathematical signs could also designate other physical concepts. Hence any sign, such as the minus sign in -1, could be understood to have a real significance, if it were taken to designate some sort of physical relation, such as position, direction, or intensity. Now of course mathematicians nowadays do not seek to introduce or justify symbolic concepts on the basis of physical considerations. Standards of justification have changed. But numerous mathematicians in the past *did* try to justify mathematical rules on the basis of physical analogies, and, moreover, some developed new methods of analysis based on physical considerations.

Anyhow, as theorists continued to ascertain new ways to interpret negative numbers in practical contexts, there yet continued to grow the multiplicity of algebraic combinations of symbols that escaped any physical interpretation. Even some elementary expressions, such as -2×-2, $1 \div -1$, $\sqrt{-1}$, and $\log(-1)$, presented interpretive difficulties. What did such expressions signify? Did they point to any actual relations between real physical quantities?

By contrast, the diagrammatic signs of geometry easily conveyed *clear ideas* of elementary concepts: straight line, square, equilateral triangle, circle, etc. The methods of geometry were celebrated as the most reliable for the search of truth: "Because, first the geometer begins by defining the least term he employs, and he never uses it but in the same sense and in the same manner. From a clear idea he deduces only consequences that are clear and incontestable, which serve as steps for rising to others that are just as clear and incontestable."[50] Mathematicians and nonmathematicians alike could clearly conceive the elements of geometry, whereas some of the elements of algebra and analysis seemed ambiguous or mysterious.

In the foundations of mathematics, mysteries were quite repugnant to some individuals. Aside from algebra, many debates transpired also over the infinitesimal calculus, wherein prominent critiques were voiced, for one, by George Berkeley, a Bishop of the Church of England (at Cloyne, Ireland, from 1734 to 1753).

Berkeley was disturbed to find that many mathematicians and students apparently accepted the principles of the calculus on the basis of faith in the authority of Newton and other masters of analysis. Berkeley argued that, whereas faith is admitted in religion, it should have no place in mathematics and science.[51] Berkeley feared that, since it is easy to manipulate signs and symbols, some mathematicians might deceive themselves by employing expressions that actually implied contradictions or impossibilities, or were empty of meaning. He rejected the expectation that anyone should just submit to the authority of the masters of analysis, and denounced it as a kind of idolatry and bigotry. In his *Defense of Free-Thinking in Mathematics* of 1735, he commented:

> In my opinion the greatest men have their Prejudices. Men learn the elements of Science from others: And every learner hath a deference more or less to authority, especially the young learners, few of that kind caring to dwell long upon Principles, but inclining rather to take them upon trust: And things early admitted by repetition become familiar: And this familiarity at length passeth for Evidence. Now to me it seems, there are certain points tacitly admitted by Mathematicians, which are neither evident nor true. And such points or principles ever mixing with their reasonings do lead them into paradoxes and perplexities.[52]

Berkeley believed in the traditional concept of scientific knowledge as based on *evident* truths, so he demanded that the principles of calculus should not be taken on trust, but should be clearly intelligible to anyone. He complained that many people inferred the truth of its principles owing to the truth of conclusions derived therefrom, noting that this inference is contradictory to the laws of logic. (For even if a method entails meaningful results and useful applications, it may yet be based on arbitrary or defective notions.) Thus he objected to various seemingly contradictory concepts in calculus: quantities that are infinitely smaller than any sensible quantity, the division of things that have no magnitude, the notion of a velocity where there is no motion, the idea that a triangle can be formed in a point, and more.

Just as Berkeley could not conceive of quantities infinitely small and equated to nothing, many mathematicians could not conceive of quantities less than nothing, nor imaginary numbers being neither greater nor less than nothing. Regarding algebra, Berkeley argued that it too should be scientific, like geometry. And he noted, like MacLaurin, that when algebraic signs are applied to represent particular things one should be careful to impose restrictions consistent with the nature of such particular things, or else one might be led to error; and he inquired "whether such Error ought to be imputed to pure algebra?"[53]

Since algebra entailed expressions that were meaningful alongside expressions that seemed meaningless, and since some of its principles were by no means self-evident, it was viewed with various degrees of skepticism. Even the founders of the calculus regarded algebra with distaste: Newton called it "the analysis of bunglers in mathematics," and Leibniz called it "a mixture of good fortune and chance." To some theorists and physicists, algebra sometimes seemed to be a nonscientific means of finding truth. Despite its high effectiveness, algebra was sometimes construed not as a science, but rather as an obscure language, or as a mechanical art.

Mathematicians and scientists were not the only ones who sought evident meaning in some of the principles of algebra. Laypersons and students, of course, also encountered difficulties. For example, the famous French novelist Henri Beyle ("Stendhal") wrote an autobiographical account of his troubles as a young lad, in the late 1790s, trying to understand the rule minus times minus is plus. He complained that his teachers:

did far worse than not explain to me this difficulty (which doubt-less is explicable because it leads to truth): they explained it to me by reasons evidently less clear to they who presented it to me.

Mr. Chabert, as I pressed him, was embarrassed, repeated his *lesson*, the one precisely against which I raised objections, and concluded by having the air of telling me:

"But it's the custom; everybody admits this explanation. Euler and Lagrange, who were apparently as worthy as you, admitted it well. . . . apparently you want to single yourself out."

As for Mr. Dupuy, he treated my timid objections (timid be-cause of his emphatic tone) with a smile of arrogance verging on repugnance. . . .

I remember distinctly that, when I spoke of my difficulty about *minus times minus* to one of the *experts*, he laughed in my face; they were all more or less like Paul-Émile Teisseire, and used to learn by rote. I often watched them say, at the blackboard, at the end of their demonstrations:

"It is therefore evident," etc.

Nothing is less evident to you, I would think. But the things in question were evident to me, and which, despite the best inten-tions it was impossible to doubt.[54]

The young Henri was horrified to find that his enthusiasm for mathematics, as an instrument for finding truth, was met with the kind of hypocrisy that he loathed in religion. Henri's father owned a copy of the *Encyclopédie*, a work despised by many priests for its secular approach to knowledge, a reaction that gave pleasure to Henri as he studied d'Alembert's articles in trying to under-stand the obscure parts of mathematics. But that did not bring sufficient clarity either, and Henri eventually came to believe, more and more, that truth was most evident only when given by the ex-perience of the senses.

Gradually, as negative and imaginary expressions found more and more geometrical and physical applications, students and the-orists developed a greater trust in algebra. But be careful to notice that in such cases the algebraic rules had been preestablished and what scientists sought were ways to apply such rules meaningfully to conceivable problems. It would have been quite a different task to begin with particular contexts and devise new mathematical concepts and methods suited precisely to such contexts. It is one thing to ascribe plausible meaning to preestablished arrays of symbols. It is quite another to devise new symbolic expressions to describe physical relations that are perceived beforehand.

chapter 4

History: Meaningful and Meaningless Expressions

Despite the growing though limited utility of imaginary expressions, objections against the significance of both negatives and imaginaries continued to arise. The notion of quantities or magnitudes "less than nothing" seemed so logically unsatisfactory to some British mathematicians that it led them to doubt the validity of algebra in general. For example, Robert Simson, the most prominent British geometer of the mid-1700s and author of the definitive new edition of Euclid's *Elements*, consistently objected to algebraic reasoning, wherever avoidable, owing to his rejection of the use of negative numbers as roots of equations.[1] Some mathematicians felt that algebra suffered owing to its habitual acceptance of arithmetically meaningless and impossible operations. This sentiment was expressed well by William Frend, a fellow of Jesus College, Cambridge, who asserted in the Preface to his *Principles of Algebra* of 1796:

> [A number] submits to be taken away from another number greater than itself but to attempt to take it away from a number less than itself is ridiculous. Yet this is attempted by algebraists, who talk of a number less than nothing, of multiplying a negative number into a negative number and thus producing a positive number, of a number being imaginary. Hence they talk of two

roots to every equation of the second order, and the learner is to try which will succeed in a given equation: they talk of solving an equation, which requires two impossible roots to make it soluble: they can find out some impossible numbers, which, being multiplied together, produce unity. This is all jargon, at which common sense recoils; but, from its having been once adopted, like many other figments, it finds the most strenuous supporters among those who love to take things upon trust and hate the labour of a serious thought.[2]

Owing to his experiences teaching mathematics to university students, Frend sought to eliminate confusion by grounding algebra on a clear "mode of reasoning, to which there could be no objection." Accordingly, he advocated "the position that algebra was nothing else than the symbolical treatment of *arithmetic*, that is, of *quantities* in the strictest sense," without negative or imaginary numbers.[3]

Frend followed other mathematicians who likewise rejected such concepts. Notably, his book included an Appendix by Francis Maseres, demonstrating how to solve cubic equations without appeal to impossible quantities. Maseres complained, in particular, that the equation

$$x^3 - bx = c,$$

had been treated "with an uncommon degree of obscurity, and has been made the subject of much mysterious and fantastic reasoning, (or, perhaps, I should say, *discoursing*, since it deserves not to be called *reasoning*) concerning negative and impossible quantities." He complained that "many writers of Algebra, even of the greatest eminence," including Newton and MacLaurin, had treated such equations "with an astonishing degree of obscurity, and almost as if they had been contending with each other which should treat it more obscurely. . . ."[4]

Some mathematicians thus believed that the principles of algebra should be at least as readily understandable as the principles of geometry. It was only natural, then, to attempt to justify the use of negative and imaginary numbers, and the operations upon

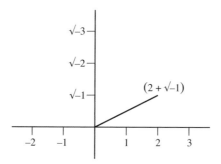

Figure 2. The sum of a positive number and an imaginary number, represented as a line.

them, in terms of some sort of geometrical constructions. Readily, negative magnitudes could be understood geometrically by supposing simply, as Descartes had done, "in a general sort of way that negative lines are directed in a sense opposite to that taken as positive."[5] Furthermore, following the work of Wallis, mathematicians had learned that a coordinate axis could consist of both positive and negative points along a line, and that two such axes could be used to plot points and figures in a plane.[6] Thus negatives had been interpreted geometrically since the mid-1600s, but it took many decades until mathematicians found a way to represent imaginary expressions in geometrical terms.

In 1797 a surveyor and mapmaker, Caspar Wessel, presented a paper to the Royal Academy of Denmark, showing a way to represent directed lines by means of imaginary expressions. Wessel sought "to avoid all impossible expressions and to explain the paradox of why sometimes one needs recourse to the impossible to solve the possible."[7] While respecting traditional mathematical rules, he found a correspondence between operations with "abstract" numbers and operations with lines. He interpreted imaginary numbers as units along a line perpendicular to an axis of positives and negatives (figure 2). Remember that Daviet de Foncenex had rejected this sort of construction as useless. But Wessel developed it very fruitfully by representing imaginary expressions as line segments (as illustrated) and by defining operations of addition, multiplication, and more, in terms of the composition of lines. Wessel argued that since his definitions were advantageous they should not be rejected. Indeed, he applied them well to the solution of problems in plane and spherical geometry. But apparently Wessel's work was ignored for almost a century. At about the same time, Carl Friedrich Gauss made similar experiments in the graphic representation of imaginary expressions, but he waited many years before making them public.

In the 1800s, prominent mathematicians, among them Lazare Carnot, Robert Woodhouse, Augustus De Morgan, and Augustin-Louis Cauchy, continued to entertain doubts as to the validity of negatives and imaginaries. This point must be stressed since, oddly, most "modern historians of mathematics have virtually ignored the fact that the question of the logical foundation of negative numbers, or lack thereof, persisted right through the middle of the nineteenth century."[8] In response, research into this issue led historian Helena Mary Pycior to assert that in the early 1800s the problem was so acute that it moved British mathematicians, in particular, "into a state of crisis."[9]

In the first decade of the 1800s, several papers on the questions of negatives and imaginaries were presented to the Royal Society of London. In 1801, Robert Woodhouse, a prominent mathematician at Cambridge, published an extensive paper defending the validity of results obtained by imaginary means. In the introductory comments to his paper, he noted that mathematicians' disagreements about algebraic quantities could undermine the semblance of logic in mathematics:

> Amongst the various objections urged against mathematical science, few oppose its evidence and logical accuracy; and, since its demonstrations have been acknowledged to proceed by a series of the strictest inferences, from the most evident principles, the study of abstract science has generally been deemed peculiarly proper to habituate the mind to just reasoning. But of late, the dissentions of mathematicians have subjected to doubt, even this "collateral and intervenient use;" for, not only has the mode of

applying analysis to physical objects been controverted, but certain parts of the pure mathematics have become the subject of dispute. Much has been heard of the science of quantity being vitiated with jargon, absurdity, and mystery, and perplexed with paradox and contradiction; so that, from the very complaints of the patrons of mathematics, its opponents may derive their most potent arguments, and abundant matter for triumphant invective.

The introduction of impossible quantities, is assigned as a great and primary cause of the evils under which mathematical science labours. During the operation of these quantities, it is said, all just reasoning is suspended, and the mind is bewildered by exhibitions that resemble the juggling tricks of mechanical dexterity.[10]

Woodhouse's paper was essentially a reply to Playfair's paper published in the same journal twenty-two years earlier. Woodhouse argued that, although Playfair's principle of analogy gave "probable arguments" and "some semblance of purpose and regularity" to the algebra of imaginary quantities, analogy was imperfect as proof and should therefore be abandoned.[11] Instead, he sought to demonstrate rigorously that conclusions drawn from the algebra of imaginary expressions followed univocally from the established rules of mathematics. In particular, Woodhouse demonstrated that impossible quantities served to represent lines pertaining to the circle, regardless of analogies to the hyperbola. He believed that the operations with impossible quantities were regulated by the rules of a logic; "a logic equally just as the logic of possible quantities."[12] Still, although he argued that the method founded on imaginary symbols was logical and easy to apply, he acknowledged that sometimes it did not abridge calculation, being neither "indispensably necessary" nor always "practically useful."

In response to the paradox that algebraic processes involving unintelligible symbols led to true conclusions, Woodhouse emphatically claimed that there could exist "neither paradoxes nor mysteries inherent and inexplicable in a system of characters of our own invention."[13] Since we invented the algebraic rules, we should be able to ascertain how seemingly confusing expressions follow necessarily from our assumptions. Thus the derivation of results would be intelligible, in *this* sense, regardless of the lack of intelligibility of basic expressions with negatives and imaginaries. Woodhouse sympathized with the ideas of George Berkeley, who argued long before that the manipulation of the algebraic symbols depends on their rules of combination rather than their meaning.

Woodhouse accepted ambiguity in the meaning of fundamental impossible expressions. He acknowledged that "it is very certain that the mind can form no idea of an abstract negative quantity; and therefore nothing can be affirmed concerning the multiplication of $-a$, and $-b$." He also commented that by itself "the ratio $1 : -1$ is perfectly unintelligible," and hence that the abstract proposition

$$a : -a :: -a : a$$

"is absurd and unintelligible and impossible to be proved." He also noted that "the symbol $\sqrt{-1}$ is beyond the power of arithmetical computation."[14] In summary, Woodhouse argued in favor of the intelligibility and truth of certain *conclusions*, obtained by means of imaginary symbols and algebraic rules, rather than of the elementary negative and imaginary quantities by themselves.

Woodhouse rejected the reforms proposed to be introduced into algebra by individuals such as Frend. He claimed that such reforms would "destroy the chief advantages of the art; its compendious and expeditious methods of calculation." He commented:

> It may, however, be justly remarked, that mathematicians, neglecting to exercise mental superintendence, are too prone to mechanical dexterity; and that some, instead of establishing the truth of conclusions on antecedent reasons, have endeavored to prop it by imperfect analogies or mere algebraic forms. On the other hand, there are mathematicians, whose zeal for just reasoning has been alarmed at a verbal absurdity; and, from a name improperly applied, or a definition incautiously given, have hurried to the precipitate conclusion, that operations with symbols of which the mind can form no idea, must necessarily be doubtful and unintelligible.[15]

Nonetheless, Woodhouse at least agreed with Frend that negatives and imaginaries should not be taught to university students until better pedagogical justification was secured.[16]

In 1802 Woodhouse presented another paper to the Royal Society, wherein he extended his original arguments. He now argued not only that algebra had a just logic, but that so "much ambiguity, confused notion, and paradox," indeed, "the evils" apparent in algebra were caused by the introduction of geometric expressions.[17] He claimed that expressions involving imaginary quantities were the "true and natural representation" of angular functions. Having already established the equivalence of such imaginary expressions to certain geometric curves, he argued that geometrical expressions were hence "not essentially necessary" and should therefore be excluded from algebraic reasoning. Otherwise, he argued, inclusion of geometric methods in algebraic analysis led to the mistaken opinion that all algebraic formulations correspond to possible geometric constructions. Moreover, this inclusion produced "confused and erroneous notions" by diverting the mind from true derivations.

Woodhouse contrasted algebra and geometry by characterizing the former as "an arbitrary system of characters" and the latter as "a system of signs, that are supposed to bear resemblance to things signified." Though geometry was effective in solving simple inquiries, he claimed that algebraic analysis provided the shortest and easiest path to truth "in abstruse and intricate" questions. He argued that the two methods were foreign to one another and should thus be kept separate. The apparent ambiguities in algebra would disappear by rejecting the "foreign" notions that gave rise to them. Nevertheless, note again that this approach justified only the methods and conclusions of algebraic analysis. It neither clarified nor justified its foundations.

In the meantime, among those who attempted to clarify the geometric significance of negative numbers, Lazare Carnot made the most extensive and systematic efforts. Carnot was an engineer and geometer who had previously won fame as a military leader of the French Revolution. He believed that the traditional theory of signed quantities was so badly founded and contradictory that he developed a new way of applying algebra to geometry based on a rejection of the concepts of negative and positive quantities. In 1801 he published *On the Correlation of Figures in Geometry*, wherein he developed an algebraic analysis of the relative positions of lines in similar figures. He reserved the word "quantity" to describe only "absolute values," such as a, in contradistinction to signed "values," such as $+a$, $-a$, $\sqrt{-a}$, so that quantities properly said would be neither positive, negative, nor imaginary.[18] Thus negative and imaginary expressions would represent not really new sorts of quantities, but geometrical relations among quantities. Then in 1803, Carnot published *Geometry of Position*, a much expanded and reworked theory for the joint analysis of magnitudes and positions of figures. This work was received well, including the good opinion of Gauss, "a good opinion not (it may be remarked) easily given."[19]

Carnot claimed that among the major questions in the foundations of mathematics, as indicated by d'Alembert, the theory of parallel lines, the notion of infinity, and the notion of negative quantities, the last remained neglected, despite its importance: "few people have attempted to analyze it in depth, though it is so deserving, all the more so, since it serves as a basis for all the operations of algebra, and that those of its notions that ordinarily are admitted without examination, are not only obscure in themselves, but false and capable of inducing error."[20] To establish the "true principles" of the algebraic theory of signs, Carnot replaced the notion of positive and negative quantities with the notion of "direct and inverse quantities." The geometry of position was thus a "theory of mutations" of geometric magnitudes, wherein the signs + and – entered only to express operations that *can* be carried out, ruling out, for example, the subtraction of a number from zero. The presence of the minus sign in the solution of a problem served only to indicate a mistake in the formulation of the problem.

At the time, many accounts claimed that negative numbers were "quantities less than zero," or that they were "opposed quantities," such as debts or coordinates to the left of zero. Carnot rejected both interpretations; the former as absurd, and the latter as vague. The historian Charles Coulston Gillispie summarized the problem: "But these were examples, not definitions, and they left difficulties. Why, for example, is it impossible to take the square root of a negative quantity? There is no difficulty in extracting the square root of a debt or a left ordinate. Why, to consider a related point, did negative quantities dominate in the multiplication of unlikes, giving their sign to the product? Moreover, there were exceptions."[21] Carnot advanced a variety of arguments against the traditional theory of signs. But for our purposes, we will mention only a few, sufficient to characterize the degree to which he objected to the traditional notions.

Carnot did not propose to change the traditional rules of operation as applied to negative numbers. Instead, it was more a question of how to interpret those rules properly by establishing a clear, systematic foundation. Still, he phrased his criticisms in rather forceful language. In his *Geometry of Position* he denounced the "falsity and absolute uselessness" of the notion of negative quantity, arguing repeatedly that "the theory is completely false," and that it "invincibly conduces to error."[22] Later, in his "Digression on the Nature of Quantities Called Negative" of 1806, he elucidated his objections. He argued that if imaginary quantities are acknowledged to be "absurd," negatives are likewise absurd quantities: "the ones and others do not differ, properly speaking, but by their different degrees of absurdity."[23]

Carnot's efforts to revise the theory of signs accompanied his simultaneous efforts to clarify the foundations of the calculus. His book, *Reflections on the Metaphysics of the Infinitesimal Calculus*, first published in 1797, became very popular as it was widely translated and reprinted for over a century. It was appreciated by leading mathematicians including Lacroix and Lagrange.[24] For its second edition, in 1813, Carnot took the opportunity to reiterate some of his arguments on the theory of signs. He argued that the ambiguities in algebra were greater than those in the infinitesimal calculus. Whereas some theorists sought to ground the operations of the calculus on ordinary algebra, Carnot deemed this approach useless, because, he argued:

> The principles of ordinary algebra are much less clear and less well established than those of the infinitesimal analysis. . . .

> . . . the metaphysics of the rule of signs is quite more difficult
> than that of quantities infinitely small: never has this rule been
> demonstrated in a satisfactory manner; it does not seem suscepti-
> ble of being so, and meanwhile it serves as base for all algebra:
> what does one win therefore by substituting it [as a basis] of in-
> finitesimal analysis? since the procedures of the first are much
> more complicated than those of the second. . . .[25]

And again he employed colorful epithets against the notions of
negatives and imaginaries. He described any isolated negative
quantity as an "unintelligible quantity," indicative of false suppositions; "a being of reason, since one cannot obtain it but by an un-
executable operation."[26]

Like Maseres, Carnot desired an exactitude of expression in
mathematical language that was generally lacking. The ways in
which mathematicians loosely defined negative numbers seemed
logically defective. Hence he distinguished casual use of language
from the exactitude that should be required in mathematics:

> In conversation, one can well say that a loss is a negative gain, be-
> cause figurative expressions are there admissible; but they are ab-
> solutely unintelligible in mathematics. Suppose that some game-
> players, seated around a table, decide that a tenth of their profits
> will be placed in a box for the poor: wouldn't one laugh at he
> who at the end of the game would claim 100 coins from the box,
> under the pretext that having made a gain of negative 1000 coins,
> he should take from the box what he should have put there if his
> gain were positive? We might tell him: we all understand that you
> have lost 1000 coins, and we're sorry for you; but the 100 coins
> must remain in the box, because your language, as clear as it is
> for describing your adventure, is not that which we use when it's
> time to count. In calculation it is necessary to call each thing by
> its name.[27]

How much the universe of meaning has changed! Nowadays, in
mathematics, figurative expressions are the norm. We are told
that "imaginary numbers" are not imaginary, that "real numbers"
are no more real than imaginary ones, "addition" and "multipli-

cation" do not necessarily imply increase, "subtraction" and "division" do not necessarily imply diminution, and so forth.

In 1813, as before, Carnot also objected to mathematicians' definitions of negative numbers as quantities less than zero, and as opposed quantities. He commented:

> To propose that an isolated negative quantity is less than 0 is to envelop the science of mathematics, which ought to be that of evidence, in an impenetrable cloud, and to engage oneself in a labyrinth of paradoxes each more bizarre than the other: to say that it is nothing but an opposed quantity is to say nothing at all; because it is then necessary to explain what is it that those quantities oppose; to recur by such explanation to new fundamental ideas resembling those of matter, time, and space, it is to declare that one regards the difficulty as unsolvable, and it is to give birth to novelties, because if one gives me for example opposed quantities, a movement towards the orient and a movement towards the occident, or a movement towards the north and a movement towards the south; I will demand what is a movement towards the north east, towards the north-west, towards the south-south-west, etc. and by what signs should such quantities be affected in calculation?[28]

Therefore, he required that isolated negative quantities "must necessarily disappear from the results of calculation, so that such results become perfectly exact and intelligible: since such are not but algebraic forms more or less implicit, and that are not susceptible of any immediate application." Similarly, he characterized imaginaries as "but algebraic forms and hieroglyphs of absurd quantities," like "chimerical beings," useful but "fictitious beings that cannot exist nor be understood."[29]

Carnot's geometrical works helped revive the pursuit of geometry after decades of neglect, eventually earning him esteem as one of the founders of the modern geometry of the 1800s. But in the long run, the critical ideas on quantity that inspired his works were dismissed. Nonetheless, his works motivated other theorists to also attempt to formulate a consistent geometric interpretation of the foundations of algebra.

In 1805, a long paper on the interpretation of negative and imaginary quantities, written by a Frenchman, the Abbé Adrien Quentin Buée, was presented to the Royal Society of London. Buée attempted to distinguish clearly between the arithmetic and geometric meanings of the + and – signs. He noted that in ordinary algebra, understood as *universal arithmetic*, these signs only meant operations to be performed on quantities. Thus agreeing with Frend and Carnot that the subtraction of a quantity from zero was absurd, he argued that a minus sign in the solution of a problem indicated that a result had to be interpreted in terms of a certain *quality*. Accordingly, he claimed that in such cases algebra did not function merely as arithmetic but as a *language*. Thus in the expression −1, the quantity 1 was a "*noun*," while the sign − was an "*adjective*."[30] He interpreted d'Alembert's claim that + and − signs indicate opposite senses as meaning that a quantity can be composed of units of opposite qualities. Likewise, he interpreted the words of Euler and others who stated that negative quantities were less than zero to refer really to questions of quality. According to such qualities, a line could have either of two opposite directions; a quantity of money could be a possession or a debt; a period of time could refer to the future or the past.

As for the sign $\sqrt{-1}$, Buée considered it not as a new sort of quantity or noun, but as a new adjective joined to the ordinary real unit noun 1. Since the sign indicated neither addition, nor subtraction, nor opposition to the + or − signs along a line, he interpreted $\sqrt{-1}$ as the "sign of PERPENDICULARITY."[31] He argued that it indicated neither an arithmetic operation nor an "arithmetic-geometric" operation, but instead a "purely geometric operation." The sign described only the direction of a line, such that if a quantity were designated by a, then its direction would be given either by $+a$, $-a$, or $a\sqrt{-1}$. Since a perpendicular direction was a quality no more abstract than any other direction, Buée denied that $\sqrt{-1}$ indicated any sort of "abstract quantity."

In addition to the geometric interpretation, Buée proposed other interpretations for the sign $\sqrt{-1}$. It could mean a sum of money that is neither possessed nor owed. It could mean a present time period, such as this year, being neither the future one, nor the past one. Yet in such interpretations he acknowledged difficulties. For example, he had to explain how the multiplication of monies that are neither possessed nor owed could result in a possession:

$$\sqrt{-1} \times -\sqrt{-1} = +1.$$

He related this difficulty to the simpler one of how a debt multiplied by a debt can give a possession:

$$-1 \times -1 = +1,$$

but he eliminated such ambiguities by reinterpreting the meaning of "multiplication," and so he claimed that such expressions were "perfectly intelligible." Nevertheless, he admitted that in some contexts, such as in pure arithmetic, the sign $\sqrt{-1}$ indicated an "absurd operation," one that cannot be carried out. Thus only when algebra was treated as dealing with pure arithmetic and the $\sqrt{-1}$ sign appeared incapable of being suppressed, only then would it describe "imaginary quantities." But in the context of algebra as a language the same sign could be meaningful and real so long as neither the quality figured by it nor the quantity affected by it were in contradiction with the conditions of the problem.

Buée's paper was a response to the widespread interpretive controversies. He spent much of his paper analyzing problems formulated by Carnot in his *Geometry of Position*. He disagreed with Daviet de Foncenex for suggesting that one should try to remove imaginary quantities from final algebraic expressions. Instead, he insisted that the sign $\sqrt{-1}$ was *necessary*.[32] Accordingly, he expressed high appreciation for Woodhouse's paper of 1801. Furthermore, he pondered the question of how the results of abstract algebraic analysis, as formulated by Lagrange and others, could be translated back into concrete physical quantities.

The same year when Buée's paper was published, 1806, a Swiss bookkeeper, Jean-Robert Argand, published a similar work.[33] Independently, Argand realized that he could represent imaginary expressions geometrically by interpreting the signs $+\sqrt{-1}$ and $-\sqrt{-1}$ as opposite lines perpendicular to the line commonly represented by the signs $+1$ and -1. He showed that any arbitrary straight line, originating from the intersection of the first two, could be represented by the expression $\pm a \pm b\sqrt{-1}$. Like Wessel, he defined the operations of addition and multiplication of directed lines in terms of joining such lines at their extremities.

Argand claimed that so-called imaginary expressions were "just as real as the primitive unit quantity," 1, and so he deemed it superfluous to distinguish any such geometric quantities as "real," "imaginary," "*impossible* and *absurd*," or "*false*."[34] He reserved the designation "truly absurd" exclusively for the notion of any quantity x that would satisfy contradictory equations, such as $x = 2$ and $x = 3$, at the same time. Indeed it was precisely from the "secret sentiment" of distinguishing the truly absurd from what is only apparently absurd that his ideas germinated.[35] In addition to avoiding the word "imaginary," Argand also avoided using the traditional signs. For example, instead of the expression $m + n\sqrt{-1}$, he used $m \sim n$, conveniently replacing four symbols with one.

Not only did Argand employ geometry to give meaning to algebraic expressions, but he introduced the basic notions of positives, negatives, and imaginaries by means of physical analogies. In the first sections of his booklet, Argand employed notions of weight, motion, balance, money, optics, and force, in addition to notions of lengths and directions, to explain concepts of quantity. Years later he was criticized for grounding algebraic expressions on such analogies.[36] In response, Argand argued that he had by no means *demonstrated*, for example, that $\sqrt{-1}$ expresses a perpendicular line, but only that he had established this meaning as a definition.[37] Already in 1806, Argand had admitted that his "principles of construction," of adding and multiplying directed lines, were based on "inductions," that is, were conceived as general rules based on specific cases. He advocated the principles of his method as "hypotheses," and argued that their validity would be established more on the basis of the exactitude of their consequences rather than on the reasonings on which they were established.[38] So here again, the principles of negatives and imaginaries still could appear as somewhat arbitrary, as definitions contrived to obtain preestablished results.

In the long run, the interpretation of imaginary numbers as perpendicular lines became widely accepted, along with its many useful consequences and applications. But in the early 1800s, some individuals continued to oppose it. For example, in 1808 the widely read critical journal *The Edinburgh Review* published a paper authored anonymously by John Playfair criticizing such interpretive attempts, and in particular, Buée's account. Playfair argued that the claim that $\sqrt{-1}$ denotes a line at 90° angles to another line was no better than saying, for instance, that it denotes a line positioned at 60° and 120° angles to the other. In the latter sense, $\sqrt{-1}$ would denote not perpendicularity but the situation that makes one of the angles twice as large as the other.[39] He praised Foncenex for having already demonstrated, "very successfully," the absurdity of the perpendicularity interpretation. Playfair described Buée as an "expert algebraist," "ingenious," "learned and acute," yet misled into "inconsistency by a kind of metaphysical reasoning." Playfair insisted that to a mathematical sign one ought not ascribe mutually contradictory meanings, such as impossibility and possibility. And he explicitly rejected Woodhouse's proposal that algebra be pursued irrespective of meaning. Now of course, Playfair still acknowledged "the power of signs," even of meaningless signs, as a useful means for discovering important and meaningful truths, and he suggested that mathematicians follow Euler in using imaginary expressions as an instrument for investigations where other more rigorous methods are too difficult to apply.

The ambiguities in the foundations of the mathematics of abstract numbers were reflected in elementary textbooks of the period. Writers of elementary texts had to wrestle with questions about the meaning of isolated negative numbers, the justification of the rules of multiplication of signs, and more. For example, Lacroix, in his widely reprinted and authoritative book, *Elements of Algebra*, explained how to reformulate problems verbally and algebraically to "rectify" absurdities and contradictions embodied by negative results.[40] In doing so, he allowed that negative quantities could serve to solve problems, at least "in a certain sense," and was not vociferously critical of prevalent notions. But not every author presented the matter so gently. For example, Thomas Simpson, author of the widely reprinted *Treatise of Algebra*, emphatically denounced prevalent approaches. In a long "Scholium" on the multiplication of signed quantities, Simpson argued:

> both $-b$ and $-c$, as they stand here independently, are as much impossible in one sense, as the imaginary surd quantities $\sqrt{-b}$ and $\sqrt{-c}$; since the sign $-$, according to the established rules of notation, shows that the quantities to which it is prefixed is to be subtracted; but, to subtract something from nothing is impossible, and the notion or supposition of a quantity less than nothing, absurd and shocking to the imagination: and certainly, if the matter be viewed in this light, it would be very ridiculous to pretend to prove, by any show of reasoning, what the product of $-b$ by $-c$, or of $\sqrt{-b}$ by $\sqrt{-c}$, must be, when we can have no idea of the value of the quantities to be multiplied.[41]

Instead of justifying the rules of the multiplication of signed quantities on any rule about isolated quantities, Simpson justified it in terms of the relations of pairs, "compound quantities." For example, instead of inventing reasons why $-2 \times -2 = 4$, one could say that since $(4 - 2) \times (5 - 2) = 6$ this result may be reproduced by establishing that

$$(4 \times 5) + (4 \times -2) + (-2 \times 5) + (-2 \times -2)$$
$$= (20) + (-8) + (-10) + (4)$$
$$= 6.$$

Here the multiplication of the "compounds" $(4 - 2)$ and $(5 - 2)$ is analyzed into four operations, including $-2 \times -2 = 4$, which jointly produce the expected result. Hence the multiplication of signed numbers is justified insofar as it serves to reproduce the results of the multiplication of subtracted pairs of numbers. Simpson acknowledged that, by contrast, other authors used analogies about positive numbers, ideas of debts, and opposites, to justify such rules, but he complained:

> But this way of arguing, however reasonable it may appear, seems to carry but very little of science in it, and to fall greatly short

of the evidence and conviction of a demonstration: nay, it even clashes with first principles, and the established rules of notation. . . .

And, farther, to reason about opposite effects, and recur to sensible objects and popular considerations, such as debtor and creditor, &c. in order to demonstrate the principles of a science whose object is abstract number, appears to me not well suited to the nature of science, and to derogate from the dignity of the subject.[42]

Simpson admitted that in *some* physical questions relating to considerations of position or opposition, negative numbers could be meaningful in the solution of problems. But in questions involving only magnitude, then negatives, like imaginaries, served instead to discover impossibility.

Other writers also acknowledged ambiguities in the treatment of negative numbers, ambiguities that entailed difficulties for students. For example, Jeremiah Day, President of Yale College, noted in his widely reprinted algebra textbook that "To one who has just entered on the study of algebra, there is generally nothing more perplexing, than the use of what are called *negative* quantities."[43] Likewise, the editor of an English translation of Euler's *Algebra*, John Farrar, commented: "It is a subject of great embarrassment and perplexity to learners to conceive how the product of two negative quantities should be positive."[44] Like Simpson, Farrar justified the multiplication of signed numbers in terms of the multiplication of pairs of numbers added or subtracted together, though noting that the operation of multiplication is used in an "enlarged sense" when negative numbers are involved. Moreover, some ambiguities persisted concerning even the results of operations on the square roots of negative numbers. For example, Day commented: "I have been unwilling to admit into the text rules of calculation which are commonly applied to imaginary quantities; as mathematicians have not yet settled the logic of the principles upon which these rules must be founded."[45] As an example, he noted that Euler and others had asserted that

$$\sqrt{-a} \times \sqrt{-a} = \pm a,$$

whereas, Day argued, the result should be not $+a$ or $-a$, but exclusively $-a$. Other writers, including Lacroix, had previously made this objection too.

One individual who invested great efforts in trying to clarify the foundations of the theory of negative numbers was the British mathematician Augustus De Morgan. Amongst his very many publications, De Morgan published several papers dealing with the significance and proper use of the + and – signs.[46] His early interest in the matter had stemmed partly from his amicable interactions with William Frend, whose daughter, Sophia, he married in 1837. De Morgan esteemed Frend as "an exceedingly clear thinker and writer," and Sophia surmised that this very clarity and directness of mind "may have caused his mathematical heresy, the rejection of negative quantities" in algebra.[47] De Morgan did not agree that negatives and imaginaries should be rejected, but nevertheless, he vigorously labored to ascertain a logical foundation for them.

In 1831 De Morgan published a book titled *On the Study and Difficulties of Mathematics*. In it, he devoted a whole chapter to a discussion of "the different misconceptions" underlying the ideas of negative and impossible quantities.[48] To give meaning to the signs + and –, when attached to single numbers or variables, De Morgan insisted that these signs designate "not quantities, but *directions*."[49] He argued that to avoid confusion,

> Above all, he [the student of algebra] must reject the definition still sometimes given of the quantity $-a$, that it is less than nothing. It is astonishing that the human intellect should ever have tolerated such an absurdity as the idea of a quantity less than nothing; above all, that the notion should have outlived the belief in judicial astrology and the existence of witches, either of which is ten thousand times more possible.[50]

To make sense of the operations with the negative sign, De Morgan recommended that the beginning student "make experience his only guide," and understand that the rules of operation "are the results of experience, not of abstract reasoning."[51] De Morgan endorsed the ordinary rules of operation of the signs, but he established them on the proviso that if a be a number greater than b then the expression $+b - a$ is "incorrect," and so he distinguished between "real" and "mistaken" forms.[52]

Like others, De Morgan knew well that negative numbers could be very useful for representing notions of position and opposition. Yet he also acknowledged that in purely quantitative contexts they were meaningless. Likewise, although he understood well the utility of imaginaries, he reiterated nonetheless the claim made by Buée (and Playfair), asserting that "the essential character of imaginary expression is to denote impossibility; and nothing can deprive them of this signification."[53] De Morgan noted that negative quantities, too, signified impossibility, at least in pure arithmetic.

Writers continued to disagree on the meaning of negatives and imaginaries. The lack of consensus may be illustrated by an example from the *Encyclopædia Britannica*. In the Preliminary Dissertations to the seventh edition, John Playfair defended "the figurative expression which gives to negative quantities the name *quantities less than nothing*," by commenting that "this phrase has been severely censured by those who forget that there are correct ideas, which correct language can hardly be made to express."[54] Yet in the same volume, another writer, John Leslie, criticized the same phrase as "very inaccurate," and proceeded to denounce such inaccurate language and ambiguous concepts in algebra:

> It is indeed the reproach of modern analysis to be clothed in such loose and figurative language, which has created mysticism, paradox, and misconception. The Algebraist, confident in the accuracy of his results, whenever they become significant, hastens through the successive steps to a conclusion, without stopping to mark the conditions and restrictions implicated in the problem. . . . Vieta, did not, therefore, discriminate the precise nature of the symbols; and his powerful example has continued to infect the language and darken the conceptions of algebraists. A disposition has also prevailed in modern times, of hastening to general conclusions, although the data be limited or imperfect. Such careless deduc-

tions are but awkwardly amended, by the adoption of expedients more like the fictions of lawyers than the reasonings of sound logicians.[55]

Leslie took the opportunity to praise Francis Maseres for his "scruples" and "aversion" against "the incorrect language and vague conception so prevalent among algebraists." Among other compliments to Maseres, he claimed, "Though not quite entitled to the rank of discoverer, that excellent person deserves a place in the history of Mathematics, for his valuable contributions and his zealous and unwearied exertions to promote accurate science."

But the idea that algebra should employ exact and meaningful language was quickly losing ground. Most mathematicians freely extended the use of traditionally well-defined words to apply to new and diverse notions. A simple example can be found in the attempts to develop geometrical accounts of the algebra of negative and imaginary numbers. Consider *A Treatise on the Geometrical Representation of the Square Roots of Negative Quantities*, an elegant book by John Warren of Cambridge University, published in 1828. Warren noted that in his *Treatise*, "whenever the word *quantity* is used, it is to be understood as signifying a *line*."[56] Ordinarily, however, one might understand "quantity" to refer to the number of units of length of a line, but Warren defined it to signify the line itself, including both length *and* direction. Also, he defined the "addition" of quantities, that is, the "*sum*" of two straight lines, as the diagonal of the parallelogram they determine (figure 3). Now notice that Warren's definitions deviated from traditional arithmetical notions. For example, supposing that line *a* has a length of 5, and *b* has a length of 4, then we would ordinarily expect that

$$a + b = 5 + 4 = 9.$$

But in Warren's account "addition" of quantities did not have this usual arithmetical meaning: the length of the line $a + b$ is *not* 9. Thus, old words and symbols for addition, subtraction, multiplication, and more began to be used in new ways. Such extensions of meaning introduced greater ambiguity into the language of

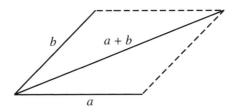

Figure 3. The geometric "sum" of two straight lines.

mathematics. But they gained currency at least because they were proposed jointly with effective innovations in symbolic analysis. Warren's work, in particular, provided a fruitful way of representing the algebra of negative and imaginary numbers in terms of geometrical constructions.

Warren's work was independent of though quite similar to the works of Wessel, Buée, and Argand. And also in 1828, another individual working independently, C. V. Mourey, proposed a similar geometric account.[57] Moreover, as mentioned earlier, Gauss too, for a long time, had been developing a geometric interpretation of imaginary expressions. Starting in 1831 he finally made his approach public.[58] Gauss used the letter i in place of $\sqrt{-1}$ (as done by Euler only occasionally), and introduced the term "complex numbers" to designate expressions of the form $a + bi$. Thus all imaginary expressions could now be construed as numbers in their own right, since d'Alembert and Euler had previously shown that any imaginary expression, however complicated, could be expressed as the sum of a real and an imaginary term.

Gauss argued that negative numbers had been accepted, because although there existed contexts in which such numbers were meaningless, there existed other contexts in which they were meaningful: "negative numbers were denied equal rights as positives, because innumerable things admit of no opposite: but the reality of negative numbers is sufficiently justified, since they find an adequate substrate in innumerable other cases. This has now been clear for a long time. . . ."[59] By contrast, imaginary numbers continued to elicit much resistance, owing to their questionable legitimacy as meaningful symbols. But Gauss argued that imaginary expressions were not a "play of symbols empty of content," because "a subject-matter just as valid can be underpinned to imaginary magnitudes as to negatives." That is, he maintained that since imaginary expressions could be interpreted as relations, they too had real meaning. Geometrically, imaginary units could be construed as located to the "*right* or *left*" of a line of positives and negatives. Note, however, that his geometrical interpretation differed from that of Warren, Argand, and others: whereas they interpreted imaginary expressions as lines, Gauss interpreted the same as points. Like Carnot, Gauss argued that the ambiguities in the theory of numbers could be solved by conceiving of so-called positives and negatives as "direct" and "inverse" magnitudes, though he went farther by designating imaginaries as "lateral" magnitudes.

In legitimizing the algebra of abstract numbers, not all mathematicians sought nor accepted geometric interpretations. Although meaning could be assigned to imaginary expressions specifically in the context of plane geometry, it was not necessarily the case that such expressions would be meaningful in all other contexts. English writers, in particular, admitted the existence of ambiguities regarding negatives and imaginaries, but continued to employ these concepts for the sake of attaining a greater, though abstract, generality in algebra. Throughout Europe other theorists displayed similar attitudes.[60] Mathematicians continued to use negative and imaginary numbers despite all uncertainties on the matter. For practical purposes, those who remained uncertain about the ultimate meaning of any negative quantity convened to employ such "merely as an algebraic symbol."[61] From this compromise emerged an increasingly abstract conception of algebra.

In response to Maseres and Frend, who sought to restrict algebra to arithmetical operations of positive numbers, the Cambridge mathematician George Peacock divided algebra into two kinds. Peacock admitted the existence of an "arithmetical algebra" in which the minus sign was used exclusively to denote subtraction of a lesser number from a greater number. But Peacock favored instead the pursuit of what he called "symbolical algebra," which included negative numbers and the operation of subtraction without restrictions.[62] Arithmetical algebra and symbolical algebra were not actually different algebras but two approaches to the same subject: the one narrower, with more restrictions, than the other. The rules of arithmetical algebra were included in the extended, more abstract algebra.

Peacock preferred to interpret algebra in the broader sense, as the science of symbols and their combinations, irrespective of any particular meaning ascribed to such symbols. He followed the approach developed earlier by Robert Woodhouse. Peacock viewed the traditional principles of algebra as ultimate and independent of any interpretation, and he denied any need to justify these principles outside of the subject, say, by any appeals to philosophy, geometry, or physical experience. In Peacock's symbolical approach, the rules of algebra appeared as arbitrary assumptions imposed by the mathematician. To establish valid rules Peacock proposed a guideline on the "permanence of equivalent forms": the *form* of algebraic relations should agree with the form of the laws of arithmetic.[63] For example, imaginary numbers were made to follow essentially the same rules as positives and negatives.

Owing to the ambiguities of negatives and imaginaries, other mathematicians, such as Augustus De Morgan, also admitted a "symbolic algebra," as "an art, not a science," determined by rules of the combination of symbols.[64] The old criterion that algebraic symbols should be meaningful, say, by representing arithmetical quantities that can be operated upon like physical quantities, was here rejected. When convenient, to altogether avoid ambiguities of meaning, mathematicians adopted the purely symbolic approach. Hence, to introduce symbolic algebra in a textbook discussion, De Morgan, for example, noted: "It is most important that the student bear in mind that, *with one exception* [the = sign], no word nor sign of arithmetic or algebra has one atom of meaning throughout this chapter, the object of which is *symbols, and their laws of combination*, giving a *symbolic algebra. . . .*"[65]

Those who shared Peacock's approach analyzed the operations and forms of algebra instead of dwelling on any questions about the meaning of the algebraic objects (magnitudes) or of their relationships. This logical approach was immensely important for the subsequent growth of algebra as a subject in itself. But this sort of approach did not appease everybody. After all, why should all branches of mathematics be expected to follow the rules of arithmetic, if the latter had scarcely been systematically justified? In retrospect, the mathematician and historian E. T. Bell commented that, "This naive faith in the self-consistency of a system founded on the blind, formal juggling of mathematical symbols may have been sublime but it was also slightly idiotic."[66]

While the approach to legitimize algebra symbolically gained popularity, so did the attempts to explain algebra geometrically. The works of Argand, Warren, and Gauss increasingly drew attention to the subject, convincing other mathematicians that imaginary expressions were susceptible of geometrical representation. Mathematicians learned that the operations of addition, subtraction, multiplication, and division could be carried out with complex numbers in terms of coordinate representations. Since complex expressions could be visualized as geometrical entities, many mathematicians accepted them as numbers, and algebra acquired greater appreciation as a consistent whole.

IMPOSSIBLE NUMBERS?

Given the old arguments and complaints against negative and imaginary numbers, we can take one of two attitudes toward the mathematicians who made such claims. On the one hand, we can choose to be unsympathetic; we can imagine that even if such mathematicians were justifiably famous owing to other achievements, they were nonetheless backward fuddy-duddies in their postures against the theory of negatives. We may thus imagine that they were merely confused, obstinate, and unreasonable. And, that despite their high-flying rhetoric, there exist good reasons underlying the theory of negatives; simple logical arguments that are now evident to us but which for some reason they could not see.

On the other hand, we can try to empathize with their objections. We may wonder whether within *their* conceptual framework, they actually had good reasons for such objections. Maybe they could see things that have since become invisible to us. Let us now pursue this latter approach, instead of assuming that their objections were simply wrong. Perhaps in forgotten old claims in forgotten old books there are still lessons waiting to be learned.

Looking back on the words of mathematicians, we can characterize the degree to which they often held positive numbers in a different standing than negatives and imaginaries.

To describe numbers such as 1, 2, 3, they used words such as "positive," "affirmative," "real," "true," "veritable," "effective," "possible," and "correct."

By contrast, to describe numbers such as -1 and $\sqrt{-1}$, they used expressions such as "negative," "imaginary," "false," "unreal," "absurd," "non-existent," "sophistic," "un-intelligible," "merely auxiliary quantities," "impossible numbers," "quantities that exist merely in the imagination," "figments," "beings of reason," "un-executable operations," "nonsense," "jargon," "incorrect forms," "mistaken forms," "mere algebraic forms," "expressions not susceptible of any immediate application," "hieroglyphs," "fictitious beings that cannot exist nor be understood," and so forth. They even appealed to expressions referring to "evil," "witches," and "judicial astrology." It was not normal for mathematicians to use such critical expressions when making a point. A reason why many of them indulged such critical language was that they were so very certain that the notions they were criticizing were so very wrong.

So how can we better understand just why such notions were allegedly wrong? One way to proceed is to consider examples that illustrate inconsistencies between mathematical notions and other notions. In particular, some mathematical rules or expressions seemed dubious because they were geometrically or physically meaningless.

Gradually, people devised interpretations that gave meaning, however limited, to such expressions. For example, already in the days of John Wallis, several geometrical interpretations gave meaning to negative and imaginary expressions. Later, some individuals, such as Wessel, Buée, and Mourey, believed that by finally finding graphic representations for certain imaginary expressions, they had found the one true significance of the square roots of negative numbers. Eventually, there came to be an abundance of interpretations. Whereas at first the square roots of negative numbers seemed to have no meaning, by the mid-1800s there were many interpretations for them in various contexts. Accordingly, De Morgan advised that when studying the works of different writers, students of mathematics should be careful to construe any claim about *the* explanation of $\sqrt{-1}$ as but one explanation among others.[67]

Unfortunately, that realization became increasingly neglected and ignored, as the interpretation of $\sqrt{-1}$ as a perpendicular line became ever more dominant and widespread. Hence, many practical men, including engineers, who understood and accepted that particular geometric interpretation, became convinced that names such as "impossible" or "imaginary" were therefore inappropriate. It was fitting that many writers generally asserted a univocal definition or meaning for the imaginary element, because that had been the tradition in defining the elements of geometry. By the 1930s, the mathematician and educator Arnold Dresden advocated that imaginary numbers could better be renamed "normal numbers," partly because in mathematics "normal" is a synonym for perpendicular, and also "in the hope that it will divest these perfectly innocent numbers of the awe-inspiring mysteriousness which has always clung to them."[68] Like some other educators and writers, Dresden claimed that "There is nothing imaginary in 'Imaginary Numbers,'" and he even asserted "the reality of imaginary numbers."[69] The name change did not stick, but some individuals at least appreciated the suggestion. One writer commented that the expression "normal numbers" carries "a much healthier sound than 'imaginary.'"[70]

The word "imaginary" seemed to have negative connotations, at least in mathematics, a field where knowledge was presumed to exist as free-standing truths, rather than as contrived inventions. At any rate, what at first had seemed to be an imaginative fiction hence became reified, and came to be regarded widely as something just as real or true as any other parcel of mathematics.

Likewise, negative numbers had seemed fictitious. Even as late as 1820, one textbook noted that if a is greater than b, then $b - a$ is not a real quantity, it is instead "an imaginary or unassignable quantity."[71] Thus negatives did not always seem real, not to everyone. But eventually, virtually all mathematicians convened to consider them a kind of "real numbers," just like positives.

Early on, some algebraic expressions thus seemed puzzling because they lacked an evident meaning. Later they gained acceptance as they were assigned some univocal meaning. Still, uncertainty about mathematical meaning proliferated not only when there was a lack of explanations but also when there was an abundance of them. Indeed, already in the 1600s there existed various interpretations of the significance of negative numbers, but ambiguities persisted. A major reason why confusion and objections persisted for so long was that mathematicians realized that some of the explanations purported for various basic rules were mutually inconsistent. Even nowadays, the physical illustrations used by teachers to explain to students the physical validity of the rules of negative numbers are often inconsistent.

For example, consider the inconsistency between common approaches to explaining the rule "minus plus minus is minus," alongside the rule "minus times minus is plus." To convince a student that

$$-4 + -4 = -8,$$

teachers say, for example, that the minus sign can represent motion to the left, so that if you move four steps to the left plus four more to the left, then you have moved eight steps to the left. This physical example agrees with the mathematical statement. But what if we apply the same concepts in multiplication? If you move *four steps to the left multiplied by four steps to the left*, how many steps have you moved and in what direction? The question seems awkward and confused. Since

$$-4 \times -4 = 16,$$

it seems that by multiplying steps to the left you must somehow change direction and move to the right. This seems odd because, by contrast, when we multiply steps to the right we don't end up moving to the left: $4 \times 4 = 16$. The physical interpretation does not correspond to the mathematics because the mathematics treats steps to the left and steps to the right asymmetrically whereas physically the directions are symmetric.

To avoid this ambiguity, teachers refine the meaning of the signs. They say that the minus sign means, more precisely, to move in the opposite direction. Hence we might interpret -4 as follows. Assume you are facing to the right and move four steps in the opposite direction, thus, you move four steps to the left. Hence, to move four times that much in the *opposite* direction is to move sixteen steps to the right: $-4 \times -4 = 16$. Perhaps you are satisfied with this interpretation.

But notice, we changed the meaning of the signs in order to make sense of the multiplication rule. What happens if we apply the new meaning to the previous example of addition? The expression $-4 + -4$ would now have a different meaning. Assume you are facing to the right and move four steps in the opposite direction so that you move four steps to the left, and move four more steps in the *opposite* direction, which is to move four steps to the right, so you move back to where you started. It now seems that $-4 + -4 = 0$. Thus the new meaning ascribed to the minus sign does not lead to the result expected in addition.

The explanation of the physical meaning of addition of negative numbers does not explain their multiplication, and the explanation of multiplication does not apply to addition. The divergence between meanings ascribed to addition and multiplication is complicated further if we compare the meaning of the quantities themselves. In -4×-4 we assigned to both minus signs the same meaning: "change to the opposite direction"; but we assigned *different* meanings to each 4: the first represents steps, the other represents how many times something is to be repeated. In $-4 + -4 = -8$, however, we gave to both minus signs the same meaning, and we gave to both quantities the same meaning, steps.

To illustrate the meaning of -4 and -4 in addition, teachers routinely give these symbols the same meaning, but to illustrate their function in multiplication, they give them different meanings, as for example, a motion in one direction and a sequence of time considered backward.

These ambiguities illustrate the sort of difficulties that mathematicians encountered when trying to provide systematic and consistent explanations for the rules of signed numbers. Analogies based on notions of quantity, money, impossibility, infinity, geometry, direction, opposition, time, negation, falsehood, repetition, motion, rotation, and more served to explain this or that rule on the operation of negative numbers, but no single such analogy served to justify all operations consistently and systematically. For example, to explain why "negative one is less than zero," many writers appealed to the notion of debt. But they disregarded this meaning when trying to justify why the square roots of negative numbers are imaginary.

Some mathematicians were aware of such pedagogical inconsistencies. In particular, the author of textbooks, Lacroix, criticized how writers usually employed inconsistent analogies when trying to justify operating with isolated negative quantities: "those who do not want to make it into an object of authority, have sought to explain the nature of such quantities, having made recourse to forced comparisons, like that of assets and debts, which are not convenient but only in particular cases. . . ."[72] Likewise, Lacroix argued that the application of algebra to geometry was not a way to conceive of the theory of negatives in its entirety, because the theory really consisted just of algebraic facts with which one should be content. He proceeded to comment that "in mathematics one abuses reason when one obstinately persists in not recognizing certain facts resulting from the combinations of calculations, which cannot be explained more clearly than by themselves." Accordingly, to avoid inconsistencies, mathematicians chose to justify the traditional rules not on any univocal physical analogies, but on abstract principles.

Even in the mid-1800s, not all ambiguities had been clarified. Looking back, one historian commented that: "The history of mathematics holds no greater surprise than the fact that complex numbers were understood, both synthetically and analytically, before negative numbers."[73]

Nowadays, people trained extensively in mathematics tend not to have any objections to negative numbers. The old "suspicion of negative numbers seems so odd to scientists and engineers today, however, simply because they are used to them and have forgotten the turmoil they went through in their grade-school years. In fact, intelligent, nontechnical adults continue to experience this turmoil. . . ."[74] To contemporary students, the notion of an imaginary number perhaps seems more peculiar than that of negative numbers. Hence, we may analyze the concept of imaginary numbers, to build a bridge toward understanding the old objections.

Consider the very name "imaginary" number. In what sense might someone rightly say that such numbers as $5i$ are imaginary? Mathematicians used to designate such numbers as "impossible quantities." Given the ordinary meaning of such words it is easy to surmise that at least in a physical sense such numbers did not correspond to anything that could be exhibited.

Consider first a simple example. Suppose a king were to approach a farmer and say: "Your harvest has been good, so in reward I will grant you more land to cultivate. *But* I shall give you a choice; you will have what you desire, just choose. What would you prefer, 4 plots of land, or $4 + 5i$ plots of land?"

Now remember, imaginary numbers are neither greater than zero, nor less than zero, nor even equal to zero. Hence, which of the choices should the farmer take? Which would *you* choose? The problem is physically meaningless; the choice cannot be exhibited in terms of actual plots of land. We only know that the two plots are apparently different in magnitude, yet it is impossible to decide which is the larger one. Therefore, perhaps, the only reasonable answer would be: "Give me the $4 + 5i$ plots of land, yes, please, if only so that I may see what you mean by such a mysterious expression!"

Thus the use of an imaginary number in this simple physical situation presents itself as an impossibility. Consider another example, this one discussed often by mathematicians before the mid-1800s.[75]

Given the number 12, cut it in two quantities that when multiplied together yield the number 36. Let's represent one of the quantities with the letter a, and accordingly the other quantity with the expression $(12 - a)$. We may thus write

$$(12 - a)a = 36.$$

By solving this equation, we have

$$0 = a^2 - 12a + 36$$

$$0 = (a - 6)^2,$$

such that,

$$a = 6$$

and

$$(12 - a) = 6.$$

So, if we cut 12 into two portions, 6 and 6, their product is 36. We can imagine this operation physically. Now, note that if instead we cut 12 into 7 and 5 their product is 35. Or if we cut 12 into 8 and 4 their product is 32. Likewise, with other attempts, we cannot seem to break the quantity 12 in a way that will yield two factors that, multiplied the one by the other, will result in a number *greater* than 36. We see no way to carry out the operation physically. But suppose somebody insists that we carry it out somehow, anyhow. Well, consider the problem algebraically. For example, let's find how we would have to cut 12 so that the product of the factors is 37. We can write

$$(12 - d)d = 37$$

$$12d - d^2 = 36 + 1$$

$$-1 = d^2 - 12d + 36$$

$$-1 = (d - 6)^2$$

$$\sqrt{-1} = d - 6,$$

such that

$$d = 6 + \sqrt{-1}$$

and

$$(12 - d) = 6 - \sqrt{-1}.$$

Suddenly we encounter imaginary quantities. The result seems odd, but we can verify it:

$$(6 - \sqrt{-1})(6 + \sqrt{-1}) = ?$$

$$(36 + 6\sqrt{-1} - 6\sqrt{-1} - \sqrt{-1}^2) = ?$$

$$(36 + 0 + 1) = 37.$$

Thus we obtain the "correct" result. Nevertheless, if we were asked *to physically split* a quantity of objects, say twelve apples, in the manner described by this algebraic procedure, we would find, surely, that we cannot. The operation is impossible. Accordingly, we can well understand why imaginary numbers were called "impossible quantities."

In 1935, the philosopher Ernest Nagel reviewed the old controversies in the theory of negative numbers as an important episode in the history of logic. In hindsight he argued:

> if "number" was nothing else than the answer to "how many" or "how much," "negative" and "imaginary" had no place in mathematics. And those who refused to admit them unquestionably had logic on their side, even though in their fierce logical fanaticism they were less wise than those who did. . . . Without a question, Frend and those like him were in the right in rejecting "impossible numbers," *if* mathematics was the science of quantity.[76]

Since mathematics indeed was deemed to be "the science of quantity," those who criticized notions of negatives and imaginaries were justified in their objections. The way in which mathematicians on the whole circumvented such ambiguities was *by redefin-*

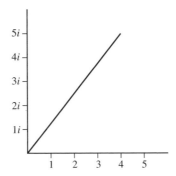

Figure 4. The complex number $4 + 5i$, represented as a diagonal line.

ing the nature of mathematics. They made arithmetic deal with
something far more abstract than quantity, namely, numbers. And
many freely reconceived algebra to deal with something that could
be even more meaningless: symbols, or "arbitrary characters."

Thus far we have considered examples illustrating the arithmetical and physical ambiguity of imaginary numbers. But for
mathematicians who for centuries had deemed geometry the most
indubitable field of knowledge, the question of the impossibility
of certain magnitudes was mainly geometrical. Accordingly, consider now how imaginary quantities could be said to be geometrically impossible.

Nowadays, owing to the works of some mathematicians during
the early 1800s, beginning students of algebra are quickly taught
ways to interpret imaginary and complex numbers geometrically.
For example, the complex number $4 + 5i$ may be represented by
the line shown in figure 4. This mode of representation served
well as a means for illustrating geometrically how operations with
complex numbers are carried out. It also led to very many useful
applications in physics and engineering. We need not rehearse
such demonstrations here. What matters now is that, whereas
mathematicians did find ways to make some sense of imaginary
numbers geometrically, there were other ways in which such numbers did *not* make sense geometrically. Although teachers emphasize the ways in which imaginary expressions are used meaning-

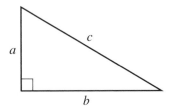

Figure 5. A rectangular triangle.

fully, there are also contexts in which they serve to signify impossible constructions. Thus, "Associating the appearance of imaginary numbers with the physically impossible is a routine concept to a modern engineer or physicist."[77] Nonetheless, since we seldom focus on exhibiting this function of imaginary expressions, it is instructive to give a simple illustration. Consider the following.[78]

Construct a right-angled triangle, and designate the lengths of its sides by a, b, c, where c is the hypotenuse, as illustrated in figure 5. Suppose that the length of the hypotenuse is five, that is, $c = 5$, and that the length of the base $b = 4$. We ask now, what is the length of a? According to the Pythagorean theorem, we expect that $a^2 + b^2 = c^2$, so that therefore, to find a we write

$$a = \sqrt{c^2 - b^2} = \sqrt{5^2 - 4^2}$$

$$a = \sqrt{25 - 16}$$

$$a = 3.$$

This result seems straightforward, clear, and true. But consider now a simple variation of the problem. Let the base now have a length of six, while the hypotenuse retains its length of five. Again, to find a, we proceed as before:

$$a = \sqrt{5^2 - 6^2}$$

$$a = \sqrt{25 - 36}$$

$$a = \sqrt{-11}$$

$$a = \sqrt{11}i.$$

But what does this mean? How can one of the sides have an imaginary length? Try to construct this triangle; try to draw it. Go ahead, try. It cannot be done.

The problem is simply that in our formulation we have violated one of the defining traits of what constitutes a right-angled triangle. We have assumed that the base of the triangle is longer than its hypotenuse, contrary to what we expect of any right-angled triangle. Hence, we cannot construct the triangle; it has no geometrical meaning. Nor can we exhibit it physically. Nor can we express the length of its side *a* arithmetically, that is, as a single number. We can thus understand why mathematicians designated such "imaginary" magnitudes as *impossible*.

Let us call this abstract and paradoxical creature a "short-hypotenuse" triangle. Whereas traditional geometry, arithmetic, and physics do not admit such triangles, algebra, however, allows us to analyze them symbolically. There is no algebraic proviso stating that the equation $a^2 + b^2 = c^2$ only has meaning when c is greater than a and greater than b. At least not in algebra as it has come to be defined since the mid-1800s. By contrast, the definition of what constitutes possible algebraic problems was still an undecided issue before that time.

For centuries, many mathematicians believed that algebraic statements were true if they could be formulated geometrically. But as algebra came to include more propositions that lacked geometrical meaning, mathematicians struggled to make sense of this divergence of two methods that they had presumed to be equivalent. In this regard, John Playfair argued:

> The cause of this diversity, in sciences which have the same object, must no doubt be sought for in the different modes which they employ to express our ideas. In geometry every magnitude is represented by a line, and angles by an angle. The genus is always signified by the individual, and a general idea by one of the particulars which fall under it. By this means all contradiction is avoided, and the geometer is never permitted to reason about the relation of things which do not exist, or cannot be exhibited. In algebra again every magnitude being denoted by an artificial symbol, to which it has no resemblance, is liable, on some occasions,

to be neglected, while the symbol may become the sole object of attention. It is not perhaps observed where the connection between them ceases to exist, and the analyst continues to reason about characters after nothing is left which they can possibly express: if then, in the end, the conclusions which hold only of the characters be transferred to the quantities themselves, obscurity and paradox must of necessity ensue.[79]

But as mathematicians abandoned the requirement that algebraic expressions represent geometric magnitudes exactly, it became superfluous to distinguish some algebraic propositions as impossible or paradoxical. Likewise, several mathematicians had advocated that algebra be restricted to a sort of universal *arithmetic*, where only such expressions as can be formulated arithmetically could be reformulated algebraically. But their restrictive approach failed to win as much support as the approach that instead sought to extend algebraic theory with ever more abstract and general considerations.

Thus algebra grew rapidly irrespective of the lack of significance of various peculiar formulations. The interpretive attitudes that facilitated this process can be characterized by the following caricatures. For example, first, suppose someone complains that short-hypotenuse triangles are algebraic abstractions and are thus "merely imaginary." In response, someone else might argue that actually other triangles are imaginary too because, apparently, there are no material objects having exactly triangular shapes. Hence it would seem that "*all* triangles are imaginary." Thus the difference between short-hypotenuse triangles and other triangles would be obscured. The difference might also be obscured by arguing in the opposite direction. Again, suppose that someone objects that short-hypotenuse triangles do not exist. In turn, someone else might contend that they *do* exist, and that their existence is proven by the laws of algebra. And, that if geometry and other traditional modes of understanding fail to describe this sort of triangle, then this is merely indicative of their inadequacy to convey certain higher truths. Moreover, they might argue that such triangles exist not only in the imagination, but as transcendent physical realities; that such figures lie midway between our plane of existence and an unperceived dimension. Though these are indeed caricatures, they are nonetheless reminiscent of other developments in the history and philosophy of mathematics that need not be listed here.

So what should we do if now we know of the "existence" of short-hypotenuse triangles? We could, perhaps, damn our ordinary faculties of understanding as being flawed. We might reject common sense. We might renounce the requirement that our geometrical explanations should be susceptible to pictorial representation. We might renounce the notion that geometrical figures describe possible material structures. We might renounce the ability of language to properly express mathematical truths. We could, in essence, redefine what it is to "understand" something, such that only those of us who *accept* short-hypotenuse triangles can be said to have understood them. And thus we might pat ourselves on the back for being able to transcend the ordinary modes of our understanding. We might then go on to form societies where we get together to discuss the properties of short-hypotenuse triangles, and celebrate one another for our abilities to go beyond the trappings of ordinary reason. We can choose to believe, in short, that short-hypotenuse triangles are just as true as $2 + 2 = 4$.

But no, whatever the happy and fun merits of such a creative approach might be, it will not be the approach followed in this book. For there are alternatives. There are other things that we can do with our time.

chapter 5

History: Making Radically New Mathematics

The demonstrations that the elements of algebra were consistent with plane geometry did not satisfy all those who desired clarification of the foundations of algebra. Notably, Gauss was not satisfied with the geometrical representation of complex numbers as a justification for them.[1] For many years he had analyzed the foundations of geometry itself. As early as the 1790s, Gauss imagined geometries different from that which describes physical space. In 1799, having failed to convince himself of the logical necessity of traditional geometric concepts, he began to express doubts about the truth of traditional geometry. By 1813 he began to formulate a new geometry systematically but without making it public. Similarly, several years later two other individuals, Nikolai Ivanovich Lobatchevsky and János Bólyai, also independently formulated new geometries. Just as traditional geometry seemed to entail no logical contradictions, neither did the new geometries. Hence these mathematicians demonstrated by example that fundamental concepts could be modified or abandoned and replaced by others. In particular, they rejected the traditional fundamental concept of parallel lines and yet constructed coherent geometries. After centuries of faith in the truth of traditional geometry, finally it began to become evident that this faith had been exaggerated.

Given this newfound arbitrariness of traditional geometry, any attempts to justify other branches of mathematics by relation to traditional geometry would thereafter lack some power to convince. Since different geometries could exist, then the correlation of operations with complex numbers to aspects of one particular geometry could hardly serve as a univocal justification for such operations. This realization, gradually, added to the reservations that many mathematicians already had about using geometric arguments to validate propositions of arithmetic and algebra. The goal of formulating algebra and arithmetic independently of geometry, as initiated by Monge and Lagrange, continued in the hands of Julius Plücker and Augustin-Louis Cauchy, among others.[2]

Moreover, there still prevailed a lack of clarity on any plausible physical meanings of imaginary quantities. In particular, mathematicians wondered whether imaginary numbers could somehow be correlated to magnitudes in three-dimensional space. A few mathematicians entertained doubts as to whether algebra as a tool was properly suited for the analysis of space. Such a feeling had been voiced in the late 1600s by Leibniz. Unsatisfied with the complications of algebra, Leibniz sought a new symbolic method of geometrical analysis that would "express situation, angles, and motion directly."[3] Concerns such as these had led a number of individuals, including Wessel, Argand, and Gauss, to search for a new method to express length and direction symbolically in three-dimensional space. Since complex numbers consist of two terms, one real and one imaginary, and these numbers had been represented successfully in two-dimensional space, it seemed plausible to search for some sort of "triple algebra" appropriate for the analysis of three-dimensional space. Such pursuits eventually led a number of physicists to develop a new mathematical theory and to advocate the complete abandonment of traditional methods.

A major contribution to the development of a new method for the analysis of space was made by William Rowan Hamilton, astronomer and professor at Trinity College in Dublin, Ireland. Like others, Hamilton was frustrated by the lack of logical foundations of algebra. In 1826, upon studying the work of a friend, John T. Graves, on imaginary logarithms, Hamilton decided to attempt to make sense of imaginary expressions.[4] In 1828, in a letter to Graves, Hamilton expressed his feelings on the matter:

> For my own part I have always been greatly dissatisfied with the phrases, if not the reasonings, of even very eminent analysts, on a variety of subjects. . . . I have often persuaded myself that the whole analysis of infinite series, and indeed the whole logic of analysis (I mean of algebraic analysis) would be worthy of radical revisions. But it would be [right] for a person who should attempt this to go to the root of the matter, and either to discard negative quantities, or at least (if this should be impossible or unadvisable, as indeed I think it would be) to explain by strict definition, and illustrate by abundant example, the true sense and spirit of the reasonings in which they are used. An algebraist who should thus clear away the metaphysical stumbling-blocks that beset the entrance of analysis, without sacrificing those concise and powerful methods which constitute its essence and its value, would perform a useful work and deserve well of Science.[5]

Several years later, in a paper drafted in 1833 and published in 1837, Hamilton openly commented "that confusions of thought, and errors of reasoning, still darken the beginnings of Algebra, is the earnest and just complaint of sober and thoughtful men, who in a spirit of love and honour have studied Algebraic Science."[6] Accordingly, he then aimed to improve the "doubtful" principles of algebra by attempting to remove "imperfections," or "confusions of thought, and obscurities or errors of reasoning." Like Playfair and others, Hamilton emphasized the contrast between the certainty of geometry and the ambiguities of algebra. He argued as follows:

> For it has not fared with the principles of Algebra as with the principles of Geometry. No candid and intelligent person can doubt the truth of the chief properties of *Parallel Lines*, as set forth by EUCLID in his Elements, two thousand years ago; though he may well desire to see them treated in a clearer and better method. The doctrine involves no obscurity nor confusion of thought, and leaves in the mind no reasonable ground for doubt,

although ingenuity may usefully be exercised in improving the plan of the argument. But it requires no peculiar scepticism to doubt, or even to disbelieve, the doctrine of Negatives and Imaginaries, when set forth (as it has commonly been) with principles like these: that a *greater magnitude may be subtracted from a less*, and that the remainder is *less than nothing*; that *two negative numbers*, or numbers denoting magnitudes each less than nothing, may be *multiplied* the one by the other, and that the product will be a *positive* number, or a number denoting a magnitude greater than nothing; and that although the *square* of a number, or the product obtained by multiplying that number by itself, is therefore *always positive*, whether the number be positive or negative, yet that numbers, called *imaginary*, can be found or conceived or determined, and operated on by all the rules of positive and negative numbers, as if they were subject to those rules, *although they have negative squares*, and must therefore be supposed to be themselves neither positive nor negative, nor yet null numbers, so that the magnitudes which they are supposed to denote can neither be greater than nothing, nor less than nothing, nor even equal to nothing. It must be hard to found a SCIENCE on such grounds as these. . . .[7]

Regarding this passage, Cyrus Colton MacDuffee commented, in 1945, that "It is startling to notice that these lines of Hamilton, written one hundred and eleven years ago, summarize so perfectly the confusion which quite generally still persists in the teaching of elementary algebra in our secondary schools."[8]

Hamilton sought an account that would give true meaning to algebra. He had been especially impressed by the controversial writings of Bishop Berkeley. Meanwhile, mathematicians led by George Peacock treated algebra as the manipulation of symbols independent of any specific interpretation. Hamilton was disgusted by this approach. He believed that algebra should not be reduced to a system of meaningless symbols and arbitrary rules. If so, algebra would not be "a Science." Instead, he believed that numbers and algebraic symbols should represent *real things*.[9] Algebraic truth had to reside in the reference to the things signified by symbols and operations. He believed that only thus could algebra be properly deemed a science like geometry, deduced from valid principles and reasoning.[10]

While British mathematicians analyzed the foundations of algebra, Hamilton pursued this sort of investigation by focusing on the concept of number. The concept of negative numbers was especially problematic because it was not easily interpreted in fundamental terms. Mathematicians defined arithmetic and algebra as the sciences of quantity or magnitudes in general, but it seemed meaningless to Hamilton to conceive of "negative magnitudes." Likewise, the derived concepts of imaginary and complex numbers seemed incomprehensible. One way to resolve this situation was to ascertain an interpretation for such mysterious symbols that would elucidate the foundations of arithmetic and algebra. Once a clear-cut or "correct" significance was established for such symbols, then the operations on such symbols would be determined by reference to the meaning of such symbols. Perhaps a way to secure meaning for negative numbers would have been to establish a direct correspondence between these and some sort of physical relations. By contrast, Hamilton chose a more abstract approach: he attempted to substantiate the concept of numbers in terms of metaphysical foundations.

At first, Hamilton was concerned with algebra neither merely as a practical tool nor as a language in need of clarification, but as an abstract system suffering from errors and imperfections of reasoning. By 1827 he believed that such problems could be corrected by reconceiving algebra as the science of time, in a similar way as geometry was understood as the science of space.[11] Previously, the philosopher Immanuel Kant had advanced the idea that the concepts of number and arithmetic originate from the purely intuitive notion of time.[12] Hamilton shared this view.

The notion that events have a specific order in time served as a foundation for the ordinal character of real numbers. Hamilton interpreted negative numbers as corresponding to steps backward in time. To avoid the ambiguous notions of imaginary expressions, he defined complex numbers as ordered pairs of real numbers, "couples," corresponding to pairs of moments of time. Hamilton argued that his "Theory of Couples" showed that "expressions which seem according to common views to be merely symbolical, and quite incapable of being interpreted, may pass into the world of thoughts, and acquire reality and significance, if Algebra be viewed as not a mere Art or Language, but as the Science of Pure Time."[13] He likewise construed the major advances in the history of analysis, "the most remarkable discoveries," those of Newton, Lagrange, and others, as stemming from the notion of time.[14]

Hamilton's formal analysis of complex numbers was received well, but his attempt to ground algebra on the intuition of time was dismissed by most mathematicians. Many already sympathized with geometric or purely symbolic approaches. Hamilton himself had been especially influenced by John Warren's book of 1828.[15] But since he did not appeal to any such geometrical interpretation of complex numbers, "it seems probable that Hamilton believed (like Gauss) that the geometrical representation was an aid to intuition, but not a satisfactory justification for complex numbers."[16]

In any case, Hamilton was not satisfied fully with his theory of couples. He desired a comprehensive system to integrate geometry and algebra. Though he had not involved geometry in his analysis of algebra of 1837, he believed that the "sciences of Space and Time" were "intimately intertwined and indissolubly connected with each other."[17] Traditional coordinate geometry did not integrate these sciences satisfactorily, especially for physics. And complex numbers could be used to represent constructions in two-dimensional space only. Warren did not write on the extension of his system to three-dimensional space, but Hamilton attempted to do so along lines pursued also by Gauss: to search for a "triple algebra" analogous to the "double algebra" of complex numbers. He tried to create a new symbolic calculus for the analysis of space by struggling to find three-part numbers analogous to complex numbers.

It was well known that a complex number such as $a + b\sqrt{-1}$ could be represented as a point or a line in a rectangular pair of axes, such that the real and imaginary parts of this number would correspond to x and y coordinates, respectively, in this two-dimensional space. So one could express the complex number as the couple (a, b) or the more typical $x + iy$. Accordingly, Hamilton speculated that for three-dimensional space, a "triplet" might have the form

$$x + iy + jz.$$

As with the usual imaginary term i, Hamilton supposed that the new imaginary term j would also have the property that $j^2 = -1$. Hamilton hoped that the triplets would serve for the analysis of lines in space. In particular, he expected that operations on such triplets would produce other triplets, just as operations with complex numbers produce other complex numbers. Sure enough, addition and subtraction of triplets resulted in other triplets.

But multiplication was problematic.[18] For example, if you multiply the triplet above by itself, you easily obtain

$$x^2 + xiy + xjz + iyx + (-y^2) + ijyz + jzx + jiyz + (-z^2),$$

which simplifies to what Hamilton wrote,

$$x^2 - y^2 - z^2 + 2ixy + 2jxz + 2ijyz.$$

Is this expression a triplet? Its first three terms are all real numbers (positive or negative or zero) and thus their sum corresponds nicely with the first term of a triplet. The fourth term corresponds also to the usual kind of imaginary term i, and the fifth term corresponds to Hamilton's new imaginary term j in his triplets. But what about the last term, $2ijyz$? It seemed ambiguous. If only it were absent it would be reasonable to say that the multiplication of triplets produced another triplet. Hence Hamilton proceeded to tinker with the rules, to try to modify them. He readily realized that one way to eliminate the term in question was to posit a new assumption: that $ij = 0$. He also realized that another way to eliminate it would be to assume instead that $ij = -ji$. In that case, in the sum of nine terms above, the terms $ijyz$ and $jiyz$ would cancel out, eliminating the ambiguous term $2ijyz$. Hence Hamilton pondered two convenient assumptions that he could posit in order to make the multiplication of a triplet by itself produce another triplet.

But a problem remained when he tried to multiply different

triplets in this manner. Hamilton expected that the multiplication of two arbitrary lines should produce a line having a length that is the numerical product of the lengths of the two lines. But his triplets, when allowed to not be in the same plane, did not have this property. Instead, he again obtained an extra ambiguous term in an equation.[19] Now, how to eliminate it? Hamilton went back to inspecting his assumptions, to see whether he could modify them in some way that eliminated the superfluous term.

Hamilton then realized that if only ij had not dropped out earlier then the equation might work. He saw that he could modify his supposed rules so that ij would not equal zero. Hence in 1843, he suddenly posited that the product of i and j should be a *new* kind of imaginary number, namely, k.

Thus Hamilton failed to find any such triplets upon which he could apply all the ordinary rules of algebra. However, he finally devised a system that satisfied most of the algebraic rules that he deemed indispensable. Surprisingly, the new "numbers" did not consist of three parts, but of four; hence Hamilton named them "quaternions."

Hamilton conceived of quaternions as hypercomplex numbers of the form

$$w + ix + jy + kz.$$

Here, not only the term i, but also the terms j and k, were deemed imaginary. Upon such numbers, he defined operations that were consistent, but did not follow all of the traditional laws of algebra. Hamilton established that

$$ii = jj = kk = ijk = -1,$$

in accordance with the ordinary rule for squaring imaginary numbers.[20] But the single outstanding deviation from ordinary algebra was that quaternions did not obey a traditional fundamental rule of multiplication. Mathematicians had believed that the product of any two numbers is the same regardless of the order in which they are multiplied:

$$ab = ba.$$

By contrast, Hamilton abandoned this "commutative" rule to multiply quaternions. Instead, he made the terms i, j, k obey the following laws:

$$ij = k, \qquad jk = i, \qquad ki = j,$$

$$ji = -k, \qquad kj = -i, \qquad ik = -j.$$

Here, the results of multiplication vary depending on the order in which terms are multiplied; $ab \neq ba$. Furthermore, notwithstanding the violation of the commutative rule for multiplication, the other traditional rules of arithmetic applied for quaternions as for other numbers.

Regardless of its similarities to ordinary algebra, Hamilton's new number system constituted a radical departure from traditional mathematics. By contrast, other algebraists, such as Peacock, had labored under the assumption that the rules of symbolic algebra should follow the rules of arithmetical algebra. Yet Hamilton discovered that the so-called laws of algebra *can be altered* to establish different systems that are internally consistent. In effect, Hamilton had developed a new algebra. Thereafter it became evident that ordinary algebra was not unique. Soon, other mathematicians, including De Morgan and Graves, introduced a variety of other new algebras.

But before such other mathematicians began to develop their innovative algebras, another individual had labored independently of Hamilton and simultaneously. A schoolteacher in the Prussian city of Stettin (or Szczecin, now part of Poland), Hermann Günther Grassmann conducted unusual experiments with algebraic signs. Grassmann's investigations began when he was pondering the role of negative numbers in geometry.[21] He regarded the lines AB and BA as opposite magnitudes. As usual, one would expect that if A, B, C are points along a straight line, such that B is between A and C, then

$$AB + BC = AC.$$

Yet Grassmann further realized that the equation would be true even if C lies between A and B. In that case, the lines AB and BC

would each extend in opposite directions. Thus Grassmann conceived of two kinds of addition: the arithmetical addition of lengths, and the geometrical addition of directed lines. He then saw that geometrical sums could be performed for any lines, even those that do not lie on one straight line.

Grassmann further realized that he could conceive of a multiplication of lines as well, by construing any parallelogram as the "product" of two directed lines. Like Hamilton, he found that if the order of the factors in such a multiplication is changed then the product varies, unless one changes its sign (that is, this multiplication was not commutative). But unlike Hamilton, he did not restrict his algebraic scheme to deal only with three-dimensional space. Also, Grassmann invented various different kinds of products. Though he had begun from geometrical considerations, he decided that his new algebra should not be limited by notions from geometry and physical experience. Hence he developed his "theory of extension" which included magnitudes having indefinitely many "orders," beyond just the three dimensions that were associated with any figure in space.

Grassmann wrote a logically detailed exposition of his ideas, as a book that was first published in 1844. But his original and highly abstract ideas found very few readers. Most mathematicians ignored his work for decades, to his chagrin. By contrast, Hamilton's work, expounded in his articles and books, attracted considerable attention, rather quickly.

Still, Hamilton's new algebra was not accepted easily. Quaternions seemed to be extremely abstract. Mathematicians already had struggled to understand imaginary expressions for centuries. The square root of negative numbers had puzzled students and experts alike. Now Hamilton's attempt to clarify the matter resulted in the introduction of two more imaginary terms: j and k. At that time, some mathematicians still hoped to dispense with complex expressions. Most prominently, Augustin-Louis Cauchy formulated a theory that would justify or replicate operations with complex numbers but without using the sign $\sqrt{-1}$. In a work of 1847, Cauchy argued that it was an expression that "we can repudiate entirely and that we can abandon without regret because one does not know what this purported sign signifies nor what sense one should ascribe to it."[22] Hence he formulated an algebra that reproduced all the algebraic properties of complex numbers and operations, but without using such numbers, nor even Hamilton's "couples."

At that very time, Hamilton was trying to gain favor for his extended theory of imaginaries. Algebraic geometry already had been made quite abstract by the efforts of Monge, Lagrange, Lacroix, and others who sought to free it from any reliance on diagrams. Then Hamilton and Grassmann introduced new and greater levels of abstraction. To mathematicians it seemed extremely unnatural to imagine that any numbers could possibly violate the ordinary commutative law for multiplication. Quaternions emerged as the most abstract extension of the concept of number ever conceived in the history of mathematics. Despite Hamilton's original interest in developing a simple method for the analysis of three-dimensional space, the concept of quaternion was difficult to interpret geometrically. This criticism was leveled at Hamilton publicly by George Airy in 1847, at a meeting of the British Association in Oxford. Hamilton himself had been disturbed by this problem soon after his invention of quaternions.

To represent quaternions geometrically, Hamilton eventually proposed that the three imaginary terms of any quaternion be interpreted in analogy to mutually perpendicular lines in space, "vectors," where the "sum" of these was itself a vector. The fourth real-number part of quaternions he construed as corresponding to a *scale* of values in *time*. Thus Hamilton introduced the concept of a "scalar" as a line that is outside of three-dimensional space, a line without any direction in space.[23] This concept could be used to characterize any numerical magnitude, such as ordinary positive and negative numbers, that do not specify any particular orientation in space. But Hamilton's early interpretation of the real part of a quaternion as a one-dimensional scale hardly clarified how this term could be *added* to the imaginary terms representing vectors in three-dimensional space. In geometry, mathematicians did not allow the addition of quantities of different dimensions; for example, it seemed meaningless to add a line to a volume, or a number to a line. Accordingly, the sum of a real number w and the expression $ix + jy + kz$ seemed geometrically meaningless to Hamilton's contemporaries. This same problem had been encountered already with complex numbers, $a + ib$, a problem that Hamilton had circumvented by writing complex numbers as "couples" instead of sums. In addition, whereas the representation of the two terms of complex numbers with two-dimensional lines on a plane seemed straightforward, the four terms of a quaternion could not be interpreted easily as a three-dimensional configuration.

Such problems of geometrical representation led Hamilton to sympathize somewhat with Peacock's symbolical approach to algebra. But he persisted in attempting to secure a geometrical interpretation, for instance, by conceiving of quaternions as the quotient of two lines, in a paper of 1849 titled "On Symbolical Geometry," and later in his *Lectures on Quaternions* of 1853. In the meantime, and even thereafter, physicists and mathematicians did not easily grasp geometrical meaning in quaternion equations.

Despite such interpretive ambiguities, Hamilton's new algebra gradually facilitated effective tools for the representation and analysis of geometrical and physical magnitudes. The discovery that multiple and distinct algebras could be established logically led some physicists to ponder whether traditional algebraic methods could be improved for the purpose of physics.

The individual most responsible for developing Hamilton's theory as a tool for physics was Peter Guthrie Tait, of Edinburgh University. By studying Hamilton's works in the 1850s, Tait became interested in quaternions because of their utility in physical applications. He corresponded with Hamilton and pursued quaternions under his encouragement. Tait's contributions served to reorient the development of quaternion methods away from pure mathematics and toward applied science.

But to introduce new mathematical methods into physics, Tait first had to fight against the apparent circumstance that, by and large, physicists were satisfied with traditional algebraic methods. Physicists already disdained pictorial geometry, not only because algebraic methods were more productive, but also because they seemed to be logically satisfactory. The emergence of new geometries was taken to suggest that the principles of traditional geometry were based on physical experience, to the extent that it seemed that the traditional formulation of such principles lacked any rigorous and purely logical justification. By contrast, analytic geometry seemed to eliminate such uncertainty.

Since algebra served to derive equalities from other equalities, its results appeared as truisms. George Boole even believed that he could express algebraically the laws of operation of the human mind. He attempted to systematically formulate logic itself in terms of algebraic equations valid for the numbers 0 and 1, such as $x^2 = x$. He expected that the laws and processes of such an algebra would be completely identical to those of true logic, except that instead of interpreting the objects as numbers, they would be interpreted as things of thought, such as "nothing" and "universe."[24]

The growing confidence in algebra embraced also analytic geometry. The latter served as the vehicle for formulating old and new geometries both. Hence, analytic geometry was construed as a means to escape the difficulties involved in the traditional formulation of the principles of geometry. Analytic methods encompassed also the many physical theories. This high degree of effectiveness had to be confronted by anyone advocating alternative mathematical methods for physics.

But Hamilton and Tait were not alone. Early interest in adopting quaternion methods in physics was shown by the Cambridge physicist James Clerk Maxwell. Maxwell crucially helped to appraise the value of quaternions for many physicists. By 1873, the year his *Treatise on Electricity and Magnetism* was published, Maxwell's reputation as a great physicist was established. In the late 1860s Maxwell became acquainted with quaternion methods partly owing to his friendship with Tait.[25] Both men believed that the mathematical methods used in physics should describe physical processes closely as they appear in nature. Maxwell's interest in quaternions thus can be traced to his views on the function of mathematics in physics.

Hamilton's systematic treatment of quantities as consisting only of magnitude, direction, or both, appealed to Maxwell, who argued that of "the methods by which the mathematician may make his labours most useful to the student of nature, that which I think is at present most important is the systematic classification of quantities."[26] He made this claim in 1870 in an address to the Mathematical and Physical Sections of the British Association, his first public statement that included a discussion of quaternions. Subsequently, in a paper "On the Mathematical Classification of Physical Quantities," Maxwell stressed his claims: "A most important distinction was drawn by Hamilton when he divided the quantities with which he had to do into Scalar quantities, which are completely represented by one numerical quantity, and Vectors, which require three numerical quantities to define them."[27] Maxwell believed that if physicists only paid closer attention to tracing necessary distinctions between physical quantities of different sorts, then it would be easier to visualize physical processes and advance scientific understanding.

Maxwell's interest in quaternions stemmed partly from reservations about the ordinary methods of analysis. Thus, for example, he wrote to Tait that he was trying "to sow 4nion seed at Cambridge," commenting that in his opinion "Algebra is very far from o. k. after now some centuries, and diff. calc. is in a mess and $\iiint$ is equivocal at Cambridge with respect to sign. We put down everything, payments, debts, receipts, cash credit, in a row or column, and trust to good sense in totting up."[28] These words reveal that Maxwell viewed algebra and calculus with significant dissatisfaction. Quaternions appeared to him a promising alternative. Maxwell chose to employ the notation of quaternions, "convinced" that the "ideas, as distinguished from the operations and methods of Quaternions, will be of great use" in science in general and in electromagnetic theory in particular.[29]

So Maxwell was not entirely satisfied with quaternion methods. One objection concerned Hamilton's rule for squaring vectors. Maxwell disliked that the square of any given vector was a negative quantity. Hamilton had established this rule in accordance with the common rules of imaginary numbers. Ordinarily,

$$i^2 = ii = -1,$$

so accordingly, when Hamilton conceived the imaginary terms of a quaternion, namely, i, j, k, he established that

$$ii = jj = kk = -1.$$

But, when vectors were considered for the purpose of physics, this rule seemed unnatural. In the expression

$$w + ix + jy + kz,$$

the terms ix, jy, kz, as well as their sum, would each be interpreted as vectors. Given any such vector, its square would equal a negative quantity. Consider a vector meant to represent a physical quantity, such as a velocity, v. The square of this vector would be negative, an odd result to interpret physically. Hence Maxwell complained, for example, that the kinetic energy of an object, $\frac{1}{2}mv^2$, would yield a negative result.[30] By contrast, in traditional algebra the square of a velocity would be positive. Owing to such ambiguities, Maxwell abstained from adopting quaternion operations fully. But he employed quaternions as a compact language for representing physical quantities and relations better than traditional mathematics.

To its leading advocates, the algebra of quaternions appeared as an extraordinary contribution to physics. To Hamilton, the advent of the new method was comparable to the invention of the calculus by Newton and Leibniz.[31] To Maxwell it was also great: "The invention of the calculus of Quaternions is a step towards the knowledge of quantities related to space which can only be compared, for its importance, with the invention of triple coordinates by Descartes."[32] To Tait quaternions were even superior to coordinate algebra.[33] Like Maxwell, Tait advocated quaternions primarily as "a Mode of Representation," ideally suited for representing physical quantities and for separating ideas of length and direction.[34] But Tait went further, by deeming quaternions as superior to any diagrams or models and claiming they served to formulate physical relations in ways simpler, more natural, and more easily intelligible than any other approach.[35] Tait construed quaternions as the best language for the analysis of space. Thus he claimed that the new method was "expressly fitted for the symmetrical evolution of truths which are usually obtained by the ordinary

Cartesian methods only after great labour of calculation, and by modes so indirect, and at first sight so purposeless, as to bewilder all but a very small class of readers."[36]

He denounced Cartesian coordinate algebra as utterly artificial and cumbersome, and argued that therefore physicists should abandon it.[37] A few other physicists agreed that quaternions embodied the culmination of the old quest for a method of space analysis.[38]

Quaternion algebra generated considerable interest among mathematicians and physicists, but still, it was used far less than its advocates hoped, especially by physicists. Very many physicists remained skeptical or uninterested in quaternions. Hamilton's algebra failed to acquire much greater currency especially because there eventually appeared a variation of it, an alternative approach, that came to be employed far more widely in physics. The new alternative was a simplified algebra that became known as vector algebra.

Before proceeding, we should note the contemporary state of the questions about the traditional rules of signs and the significance of imaginary expressions. By the 1860s, most mathematicians and educators overtly espoused the usual theory of signs. By then few cared much whether such rules had any "impossible" connotations in any particular physical contexts and were content to just claim that, if so, such rules should not be applied in such ways. The more abstract rules of algebra were increasingly taught to ever younger students irrespective of physical or intuitive meaning. The word "impossible" became increasingly rare in expositions of numerical algebra, and students became less troubled by symbolical expressions that a century earlier had been widely deemed impossible.

Incidentally, Charles Dodgson, a lecturer of mathematics at Oxford, wrote stories for children based partly on his knowledge of the expanding world of mathematical meaning. In one fictional story, published in 1872, he recounted a dialogue between a Queen and a little girl called Alice, in which the Queen asked Alice to believe something rather incredible:

> Alice laughed. "There's no use trying," she said: "one *can't* believe impossible things." "I daresay you haven't had much prac-

tice," said the Queen. "When I was your age, I always did it for a half-an-hour a day. Why, sometimes I've believed as many as six impossible things before breakfast."[39]

Despite the general acceptance of the traditional theory of signs, at the same time as new algebras were being developed, there continued to exist some objections to the standard notions of negative and imaginary quantities. For example, in a textbook of 1864, Horatio N. Robinson noted that "We cannot, numerically, take a greater quantity from a less, nor any quantity from zero, for no quantity can be *less than nothing*." Instead, he construed positive and negative signs as "qualities," and explained algebraic subtraction as consisting of a change of quality of direction.[40] Following Carnot, some authoritative texts continued to describe the subtraction of a greater number from a lesser as an "un-executable operation."[41] Some texts continued to characterize such operations, as well as the extraction of square roots of negative numbers, as impossible operations. Some writers still sought to provide logical justification for the rules of signed numbers. William Cain, for example, acknowledged that the extension of the rules of arithmetical algebra to signed numbers was usually done without rational basis, by appeals to a disconnected variety of notions, and hence he required that instead the signs + and − should be given univocal meaning.[42] By establishing narrower definitions, certain problems then seemed illusory. Thus, in response to d'Alembert and Carnot, Cain argued that the expression

$$1 : -1 :: -1 : 1$$

implied really no equality between ratios of greater and lesser quantities; he rejected such assertions as absurd extensions of arithmetical assumptions. Thus writers and educators continued to reinterpret old ambiguities, by formulating new accounts in accord with new notions of mathematical validity.

Among professional mathematicians, Leopold Kronecker began to advocate a program to radically reformulate the numerical foundations of mathematics. Kronecker sought to show that all of modern mathematical analysis could be cast exclusively in terms of relations amongst the sequence of "ordinal numbers": first, second, third, fourth, and so on. From such numbers, he argued, we can build other mathematical objects. Thus he sought to eliminate fractions, irrational numbers, the concept of continuity, and more, from the elements of analysis. He famously claimed that "God made the integers; all the rest is the work of man." Cauchy had earlier shown how to banish imaginary numbers. Likewise, following a method advanced by Gauss, Kronecker showed how to avoid or replace negative numbers by using instead certain algebraic congruences.[43] He also showed how to avoid fractions and so-called algebraic numbers. But he did not complete his program for all kinds of numbers and analytical forms. Ultimately, he wanted to banish all infinities from mathematics.

At any rate, many mathematicians found Kronecker's program extremely distasteful. By then most were quite pleased to admit negatives, irrationals, imaginaries, and so forth, as legitimate elements of mathematics, contrary to the arithmetic of the ancients. But still, a few mathematicians were sympathetic. In particular, Richard Dedekind too acknowledged and pondered "the gradual extension of the number-concept, the creation of zero, negative, fractional, irrational and complex numbers."[44]

Finally, furthermore, it is worth mentioning that at Cambridge University, even in the 1880s, imaginary expressions were still regarded as dubious by comparison to real numerical expressions. This was a characteristic of their educational system, which back then constituted part of their alleged standards of rigor. But such old practices are nowadays often dismissed as an unfortunate case of retrograde teaching techniques.

Thus, despite changing notions of the nature of mathematics, negative and imaginary numbers continued to be interpreted in some instances as signifying ambiguities or impossibilities. Accordingly, when some physicists refined the algebra of quaternions to suit their purposes, they rejected its physically meaningless aspects: negative and imaginary elements.

Vector algebra emerged from quaternions as a simplification tailored to the needs of physicists. It was developed chiefly by Josiah

Willard Gibbs and Oliver Heaviside, working independently of each other and almost simultaneously. Gibbs, a professor of mathematical physics at Yale University, first became acquainted with quaternions by reading Maxwell's *Treatise*, and decided to pursue Hamilton's methods to master electromagnetic theory. Meanwhile, Heaviside, a retired telegraph operator in England, also studied quaternions owing to Maxwell's *Treatise*. Early on, Heaviside realized that operations with quaternions did not yield results in full agreement with traditional mathematics, and he wished the two would harmonize better. Also, the quaternion approach did not seem sufficiently convenient when applied to electrical theory. Heaviside appreciated the utility of vectors, as Maxwell and Tait had employed them, but he saw no need to justify the rules of vectors in terms of quaternions.[45] He complained, for example, that "some of the properties of vectors professedly proved were wholly incomprehensible. How could the square of a vector be negative?"[46] He characterized the negative sign in the square of any vector as "the root of the evil."[47]

Quaternions seemed to hinder physical understanding and obscure the rules of vectors. In accord with Maxwell's criticisms, Gibbs and Heaviside modified the principles of vectors to increase their utility. They realized that although Hamilton and his followers stressed the utility of quaternions in physics, in actuality only vectors were involved in most applications.[48] For Hamilton and Tait the vector

$$ix + jy + kz$$

was itself a quaternion,

$$w + ix + jy + kz,$$

where $w = 0$. By contrast, Gibbs and Heaviside saw no practical need to mix the "imaginary" expression with any real number w. This mixture went against Maxwell's prescription to distinguish between physical quantities. Therefore, they posited the vector as the fundamental concept.

Furthermore, Gibbs and Heaviside simplified the notation of vectors *by eliminating all references to imaginaries*. They freely changed Hamilton's rules to establish that the square of any vector be positive instead of negative. Thus they abandoned the original association of vectors and imaginary numbers. Moreover, instead of basing the analysis of space on the quaternion concept, or on the terms i, j, k, or on triple coordinates, Gibbs and Heaviside based it on vectors represented simply as *single* characters. And they entirely excluded quaternions.

Gibbs' *Elements of Vector Analysis* owed much to the works of quaternion theorists, but his new notation facilitated the adoption of vectors by many physicists. Gibbs appealed to physicists who, like Maxwell, sympathized with quaternion ideas but found quaternion methods unsatisfactory. Gibbs demonstrated how vectors simplified physical problems, facilitating methods for their solution.[49] Heaviside, too, derived his system from that of Hamilton and Tait, "by elimination and simplification," to devise "a thoroughly practical system."[50] He described it as follows: "It is simply the elements of Quaternions without the quaternions, with the notation simplified to the uttermost, and with the very inconvenient *minus* sign before scalar products done away with."[51] In 1893, Heaviside published an introduction to vector algebra in the first volume of his *Electromagnetic Theory*.[52]

Some of the advocates of quaternions rejected vector algebra as a monstrous hybrid. For example, the Scottish mathematician Cargill Gilston Knott complained that by making the square of a unit vector non-negative, the proponents of vector theory were abandoning the usual associative law of multiplication.[53] In Hamilton's algebra the association of terms does not affect the result:

$$i \times (j \times j) = (i \times j) \times j.$$

The same rule was also true in traditional algebra. However, in the Gibbs-Heaviside algebra,

$$i \times (j \times j) \neq (i \times j) \times j,$$

because

$$i \times (0) \neq (k) \times j$$

$$0 \neq -i.$$

Thus, advocates of quaternions used arguments about algebraic elegance against vector theorists, just as others had previously criticized Hamilton's abandonment of the commutative law. By the 1890s, however, there existed many different algebras, so the departures of Gibbs and Heaviside from traditional rules could just as well be accepted, especially in the practical interest of physical mathematics.

Since many physicists disdained quaternions, Heaviside and Gibbs strategically dissociated their system from that of Hamilton.[54] Vector theorists argued that their method should be adopted alongside the methods of coordinate algebra and that the two were equivalent. Heaviside presented vector mathematics as a "language" equivalent to Cartesian analysis but better suited to represent physical magnitudes with "natural" simplicity and compactness.[55] He complained that "in the Cartesian method, we are led away from the physical relations that it is so desirable to bear in mind, to the working out of mathematical exercises upon the components. It becomes, or tends to become, blind mathematics."[56] Critical of the "bulky inanimateness" of Cartesian coordinate algebra, Heaviside argued that symbolic expressions should represent their physical counterparts clearly, such that "the mere sight of the arrangement of symbols should call up an immediate picture of the physics symbolised."[57] Gibbs, likewise, presented vectors as convenient abbreviations of coordinates.[58] This tactic worked, as physicists more readily incorporated the new vectors into their researches, while quaternions remained rather neglected.

In the new year 1900, the physicist August Föppl aptly reprised the problematic development of physical mathematics. In the Foreword to his *Introduction to Mechanics*, Föppl placed special emphasis on the question of the physical adequacy of mathematical methods. In his discussion, he compared the usefulness of various mathematical methods:

> It is especially noteworthy that mathematical theories, as they are carried out everywhere nowadays, are cast in a form that is not particularly favorable to their immediate application in mechanics. In mechanics, and above all, in theoretical physics, one deals

mainly with the analysis of directed magnitudes. For such an application, despite all the high esteem we must emphasize for the achievements of contemporary mathematics, the widely available articles of apprenticeship are inadequate. Algebra and Analysis involve only the relations among magnitudes without direction; the one apparent exception, the well-known geometrical interpretation of complex magnitudes, had this aim but in fact damaged more than availed the subject. To be sure, analytic geometry demonstrated how with the help of the coordinate method one can also submit directed magnitudes to calculation, but this means of information necessitated a diffuse representation that is little conducive to clear interpretation.[59]

Like other physicists, Föppl understood that mechanics essentially involved the analysis of directed magnitudes. Accordingly, he emphasized that physics required mathematical methods that would be suited ideally for this purpose. He agreed with physicists who argued that coordinate algebra failed to facilitate a clear representation of directed magnitudes. The methods advanced over the last decades had the advantage of treating geometrical relations with great generality, as well as incorporating, to some extent, the notion of directed magnitude, but such methods remained insufficiently developed for use in physics.

The development of mathematical methods ideally suited to the needs of physics had been neglected. But at least tentatively, physicists finally had an adequate mathematical instrument: vector methods. Föppl wrote:

A healthy start towards the improvement of the mathematical resources used in mechanics, presently, has been made by a few prominent mathematicians for a long while already. For the time being, they do not extend particularly far, and unfortunately, up to now they have had but slight influence upon the formulation of the prevalent mathematical theory. One must hope that the future will bring what hitherto has been neglected; many signs seem to point to this. In the meantime, we must try to manage with the mathematical means that currently have become available to us.[60]

Vector methods appeared as the best available tool for the analysis of physical magnitudes, even though they had not yet been developed extensively. Thus Föppl employed vector methods thoroughly in his textbooks on mechanics and electromagnetism. Likewise, other physicists came to rely increasingly on vector mathematics.

Despite Föppl's hope for further improvements, the fundamental mathematical methods used today in physics textbooks are essentially the same as in 1900. Thus we may end our historical excursion at this point. By then the debates about the significance of negatives and imaginaries had diminished; confusion continued to arise, nonetheless, but was relegated mostly to school classrooms where ever-younger students were taught the traditional symbolic rules.

To summarize the history of the attempts to make sense of the ambiguous algebraic expressions we may turn to the words of Augustus De Morgan. In 1849, De Morgan argued that algebra had been brought to its present state through "three distinct steps." First, people believed that unintelligible expressions were also impossible in nature. Thus, since the idea of quantities less than nothing was unintelligible, it was rejected. The second step, however, placed full trust on the results of algebra regardless of their degree of intelligibility:

> It consisted in treating the results of algebra as necessarily true, and as representing some relation or other, however inconsistent they might be with the suppositions from which they are deduced. So soon as it was shewn that a particular result had no existence as a quantity, it was permitted, by definition, to have an existence of another kind, into which no particular inquiry was made, because the symbols would give true results, [which] did not differ from those previously applied to the old ones. A symbol, the result of operations upon symbols, either meant quantity, or nothing at all; but in the latter case it was conceived to be a certain new sort of quantity, and admitted as a subject of operation, though not one of distinct conception. Thus, $1 - 2$, and

$a - (a + b)$, appeared under the name of negative quantities, or quantities less than nothing. These phrases, incongruous as they always were, maintained their ground, because they always produced a true result, whenever they produced any result at all which was intelligible: that is, the quantity less than nothing, in defiance of the common notion that all conceivable quantities are greater than nothing, and the square root of the negative quantity, an absurdity constructed upon an absurdity, always led to truths when they led back to arithmetic at all, or when the inconsistent suppositions destroyed each other. This ought to have been the most startling part of the whole process. That contradictions might occur, was no wonder; but that contradictions should uniformly, and without exception, lead to truth in algebra, and in no other species of mental occupation whatsoever, was a circumstance worthy of the name mystery.[61]

Negative numbers, at first contrived by merchants and accountants for practical purposes, had been disdained by early mathematicians as useless fictions, but then became admitted as "real numbers." Finally, the third step, results involving negative and imaginary numbers were accepted increasingly as systematic interpretations of them were devised and disseminated. Mathematicians agreed that symbolic algebra had an internal logic even if it did not always lead to meaningful results in specific applications such as physics. Thus De Morgan commented: "any of those who are still bewildered by an art in which impossible quantities, or quantities which are not quantities, are made objects of reasoning, should become aware that by slow degrees, and by the union of many heads, the art has become a science, and the impossibilities possible. . . ."[62] Here De Morgan did not mean that algebra was a science, in a physical or empirical sense, but that it was logical. The "impossibilities" became plausible at least as expressions derived from symbols according to systematic rules.

It was because of symbolical algebra that mathematics on the whole ceased to be defined as the science of quantity. The advocates of symbolical algebra argued that, although magnitudes or numerical ratios had served to ascertain the form of most mathematical structures, such structures, and others, could stand independently of such notions. For example, George Boole argued that, rather than define mathematics as the science of magnitude, the defining characteristic of any true calculus is that it be "a method resting upon the employment of Symbols, whose laws of combination are known and general, and whose results admit of a consistent interpretation."[63]

To De Morgan's characterization of the history of algebra we may add a fourth step. Upon the invention of quaternions, mathematicians realized that they had considerable freedom to invent new algebras. Any question about whether a particular symbolic expression is meaningful, possible, or true came to depend more clearly on the basic rules invented or adopted in order to decide the question. This last realization will serve to shed new light on the earlier debates about the significance of ambiguous symbolic expressions.

FROM HINDSIGHT TO CREATIVITY

Reader alert! You have reached the heart of the book! This is it, this section contains the main argument. Everything else consists essentially of examples. Having delved into history we soon will proceed to experiment with concepts and signs. So it would seem that we now stand on a bridge between hindsight and creativity. But do not rush ahead, as this is not really a bridge, it is the main destination, the important part. What follows afterward are just experiments, and at the very end there will be no bang, no punch line, no big surprise. What precedes this section simply justifies it, and what follows are merely illustrations. This is it.

So what can we learn from history?

First, of course, we can learn all the tried and true lessons about how minus times minus is plus, and about how mathematicians gradually freed algebra from old associations with ordinary experience, which, in the past, led some able thinkers astray, to apparent paradoxes and confusion. Thus negative numbers have often served as a stepping stone for students on the road to mathematical abstraction. Likewise, they have been immensely useful in countless physical applications, which we need not review here. Many books and teachers already tell the story, well enough. But meanwhile, an aspect of the history remained neglected: the connection between interpretive disagreements and formal innovations.

Major developments transpired in geometry and the calculus during the 1800s, while some writers continued to quibble and debate over the theory of negative numbers. They continued to voice old-fashioned complaints, as though stuck in confusions of the past. Did anything good come out of that? Surprisingly, yes, major advances stemmed directly from debates about the negative sign. We reviewed how four major developments thus originated: the distinction between symbolical algebra and arithmetical algebra, the formulation of the Theory of Couples, the creation of the algebra of quaternions, and the new vector algebra. The study of new algebras was thus rooted in the controversies over signed quantities. Just as the old concept of parallel lines catalyzed the formulation of new geometries, so too the old concept of quantities less than nothing played a seminal part in the birth of nontraditional algebras. The growing varieties of mathematical elements and systems contributed to heightened attention to the logic and meanings of signs. Moreover, mathematics was redefined, on the whole, so that it was no longer defined as the science of quantity.

Before the invention of new geometries, people believed that mathematics was a unique and universally true system of knowledge. They were wrong. Mathematics was no longer *one*, but many. And in view of the new geometries, physicists gradually and grudgingly acknowledged that maybe the old geometry was not necessarily the best means for the precise description of nature. Eventually, more and more physicists became convinced that the old geometry in fact was *not* the geometry of nature. This development transpired owing mainly to Albert Einstein's theory of gravity, published in 1916. Einstein formulated his theory by employing a nontraditional geometry. And astronomers soon found observational evidence that fit better with Einstein's mathematical scheme than with Newton's. The details of this dramatic development need not concern us here. What matters immediately is that, it raised awareness that, despite the long and successful use of traditional geometry, its principles do not necessarily describe nature exactly.

By analogy, we may ask, what about the traditional principles of arithmetic and algebra? Do elementary quantitative concepts and methods apply perfectly in every physical context where we employ them?

Following Hamilton's invention of quaternions, in 1843, mathematicians realized that they could invent new algebras. Soon, many other algebras were invented. Like the new geometries, the new algebras were highly abstract. Some of these algebras, especially Hamilton's theory of quaternions, were effectively applied to describe and analyze some physical phenomena. Nonetheless, many physicists remained unsatisfied with the new methods, and continued to prefer traditional forms of algebraic analysis. For the most part, the new algebras seemed exceedingly abstract for the representation of physical phenomena. The new principles, such as the notion of a four-part imaginary number, were far removed from ordinary empirical notions. Quaternions were a new element in the foundations of mathematics, that is, a new sort of number with certain seemingly strange properties. Some of the traditional "laws" that applied to other numbers did not apply to quaternions. Yet the theory did not involve the rejection of any traditional rules of algebra as applied to the older, standard numbers. Therefore, Hamilton's new algebra did not challenge directly the validity of traditional algebra. By contrast, new geometries constituted a direct conceptual challenge to the validity of traditional geometry because they were based on the rejection of some of its fundamental principles.

In retrospect, it seems ironic that geometric notions that for centuries had been deemed to be so clearly true in connection to the physical world eventually came to be viewed as not exactly valid in certain applications, whereas algebraic principles that for centuries had seemed notoriously ambiguous came to be used virtually without qualms.

The heated disputes about the significance of such elements as negative and imaginary numbers, and their rules of operation, eventually died down, as these subjects ceased to generate major concerns. It is beyond the scope of this book to explain why such concerns disappeared from the complexion of mathematics, though a few points are worth noting. The lazy answer, of course, is to assume that the old objections finally faded simply because they deserved to die; that mathematicians finally came to terms with the one true and correct theory of negatives. Surely, the major reason was that most mathematicians indeed became satisfied with the growing abundance of systematic formal justifications for the standard traditional rules. Yet other factors should also be considered.

First, we should note that as faith in the certainty of traditional geometry eroded, algebra emerged by default as an alternative on which to place one's faith in mathematical analysis. As geometry fell from grace, certain traditional modes of criticizing algebra on the basis of geometry fell out of fashion. Hence it became irrelevant to criticize any part of the theory of numbers in terms of impossibilities in geometric constructions.

Moreover, the widespread fragmentation of disciplines into "independent" conceptual domains belittled the traditional drive to make the principles of one discipline intelligible in terms of another. Thus the desire to regulate or restrict algebraic theory on the basis of geometry, arithmetic, physics, metaphysics, or other considerations, came to be seen as misguided.

We might note also that debates about negatives and imaginaries decreased as textbooks incorporating the standard rules and notions were used to instruct ever younger students. Could it be that some part of the acceptance in question transpired because students of mathematics became accustomed to the standard rules at ages too early to carry any critical doubts into later life? It is clear, at least, that teachers have had to deal with the confusion of students, and accordingly, that textbooks evolved in ways that sought to eliminate possible confusions by maximizing the apparent simplicity of the subject.

In any case, traditional elementary mathematics has been thoroughly elaborated and applied in countless physical contexts, yet we cannot assume that it constitutes the only conceivable system of symbolic representation. Time and again, theorists having diverse occupations and concerns have devised innovative mathematical methods for representing and analyzing physical relations. In certain cases, as in the creation of new geometries, and in the creation of vector theory, new mathematical systems were devised by modifying certain preestablished rules and systematically exploring the consequences.

For example, (and this is the most important historical example in this entire book) Hamilton had established the rules according to which one should operate with quaternions and vectors, but decades later Gibbs and Heaviside chose to change the rules of vector algebra in order to formulate an algebra more useful for physics. They kept most of the preestablished rules but they changed a few. In particular, they made the square of a vector be positive rather than negative, because they construed the negative sign as physically inappropriate. To do so, they had to change the rules of vector multiplication. And once the square of a vector was no longer negative, it became meaningless to conceive of vectors as imaginary numbers; at least in the new theory. Thus, by modifying the rules of quaternions, they devised a new algebra.

Now of course, the development of a new system of representation and analysis stemming partly from another does not entail that its predecessor be necessarily abandoned or superseded. Both systems can be pursued, even independently of one another. And so it was with quaternion and vector algebras, both of which continued to be pursued in accord with the interests of mathematicians and physicists. Quaternions received more attention from mathematicians, and vectors were employed more by physicists, though quaternions too were used in various applications. Furthermore, quaternions and ordinary vectors received less attention from mathematicians than was given to other systems that were, by contrast, not restricted to three-dimensional space.

Hence, notwithstanding the successful efforts to ascertain logical foundations for traditional symbolic procedures, we may yet analyze the extent to which such procedures serve to represent physical operations. Mathematicians rigorously study the formal consequences of whatever set of symbolic rules they choose to investigate. Often they investigate formal systems that have greater generality. Likewise, we should carefully study the symbolic and procedural methods we employ as tools of physical representation and analysis. The student of nature may well look at the most general mathematical systems, but it is instructive to begin by considering simple, narrower applications of concepts to particular contexts and cases.

In this process, we can even reinspect the elements of arithmetic and algebra in light of old questions about meaning. We have, after all, one major advantage over mathematicians prior to the mid-1800s; namely, that now

we *know* that the principles of mathematics can be modified.

In essence, we know that it is possible to construct new theories involving rules that deviate from traditional methods. By contrast, most mathematicians who, before the 1840s, objected to standard notions believed nonetheless that the established *principles and procedures* were essentially true. And even when some of them debated not merely questions of interpretation, but also the actual results of definite operations, they yet construed their disagreements as meaning that, at most, only *one* party could be right. Had they only known then that multiple algebras can be devised, rather independently of one another, they could have presented their conflicting views as *different algebras*.

By pursuing unusual mathematical notions systematically, we can develop new mathematics. By developing mathematical rules that correspond to observed relations among things we can formulate new symbolic methods of representation. Then, questions as to whether any such algebra is valid are just questions about the internal consistency of its results with its premises. And questions about the physical validity of a particular algebra can be answered mainly on the basis of practical usefulness.

chapter 6

Math Is Rather Flexible

In the remainder of this book we will carry out various conceptual and creative experiments. The first goal is to show that there is a certain plasticity in some of the concepts we use in mathematics. The second goal is to show how this plasticity can be used to devise new tools for physical representation and analysis.

For now, we will focus on the first goal. What does it mean to say that some concepts can be modified? Does it mean that such concepts are arbitrary? Hardly. Concepts arise in connection with restrictive conditions and needs. We rarely invent concepts arbitrarily and irrespective of systematic considerations; at least not in physics or mathematics. But once we have a concept, we can proceed to refine it, extend it, or modify it. Of course, most mathematical concepts have far less plasticity than many other concepts. For example the word "art" has acquired, throughout history, many more connotations and variations of meaning than, say, the notion of "multiplication." Yet even notions such as "multiplication," "subtraction," and "number" have been refined, extended, and revised throughout the centuries.

Suppose we assign a new meaning to a traditional concept. If we wish to be silly, we might say that "multiplication" is the name of a blue tree. Here we have merely assigned an *entirely* new meaning to a familiar word. By contrast, suppose instead that we stipulate only a *partial* variation in the usual concept of multiplication, by saying, for instance, that any negative number multiplied by a negative number shall yield a negative number. Here we might perhaps say that we have "modified" the concept of multiplication, that is, introduced a variant of its usual meaning. In the new variant, 3 times 4 is still 12, and 4 times 4 is still 16, and so forth; but the multiplication of negative numbers will yield modified results. What are the formal consequences of such changes? And can they possibly be of any practical use?

To reinspect the traditional theory of negatives, for example, we need not reiterate old objections against it. But we may try to ascertain what aspects of such objections involved critical insights, or physical notions that may yet be of some use. Or, such considerations aside, one might just proceed directly to formulate alternative theories, by "modifying" traditional mathematical principles, that is, by devising distinct algebras.

Such artificial schemes might perhaps be employed as tools for physical representation. Of course, the point is best made if the alternatives offer certain advantages. So, in the interest of practical simplicity, we might try to devise examples that reduce or eliminate asymmetries and complications that exist in the traditional framework. That said, we must note, however, that it is possible also to devise theories involving new and greater complications. Moreover, and remember this: the simplicity of a system does not necessarily mean that that system constitutes better mathematics than another, nor that it will systematically serve as a better means of physical representation.

So we proceed now to illustrate simple ways to use traditional mathematics to devise alternative though rudimentary concepts and symbolic systems. In light of the preceding historical review, let us now reassess some of our notions about signed numbers. To exercise our imagination, let's begin with a simple playful example. Regarding the old debates about the incorrectness of the notion that negative numbers are less than nothing, let's consider the subject afresh. Can we make sense of the old claim, stressed even by the prominent physicist and mathematician Jean d'Alembert, that negative one is not less than zero? By pondering this question we will begin to surmise that some long-dead writers wanted mathematics to have certain basic symmetries that are obvious in physical experience, but are otherwise absent from the common rules of signs.

SOMETIMES –1 IS GREATER THAN ZERO

Yes, it sounds like nonsense. Doesn't it?

Consider the old notion of "quantities that are less than nothing." Let's try to defend the now common notion that negative numbers are numbers that are less than zero.

One reason used by past mathematicians against these notions was to say that to attain a negative number one would have to take something away from nothing, an operation that is altogether impossible and even inconceivable. In response, we can say, very well, this objection is reasonable; we agree that *quantities* cannot be less than nothing, but we disagree that this point has any relevance to the properties of negatives because negatives are *not* quantities. We nowadays define mathematics in such a way that numbers do not necessarily represent quantities, in the way that we know quantities physically, and thus the difficulty disappears.

Or does it?

A past mathematician could smile, and say, "well, at least you do *agree* with me that there are no quantities that are less than nothing, even if this seems to you a trivial point. But what now if I further specify that any *negative numb*er is not less than nothing?"

We think about this, and reply that, again, the ambiguity is merely verbal, really, that the comparison in question should concern negative numbers and zero, not negative numbers and nothing. The concept of nothing is vague, and although we can indeed often use the number zero to represent "nothing," say, in physical applications of mathematics, the two concepts are not the same. "Nothing" is not the only concept zero can represent, and hence the relations between zero and other numbers are not ruled by having to follow any properties that we associate with the concept of nothing. For example, the number zero may be used instead to represent a particular point on a scale, that is, a location on a reference frame.

Our imaginary mathematician might now reply: "Fine, we need not argue the difference between zero and nothing. But I can restate my objection, to show that it's not merely verbal, and say that negative numbers are not less than zero."

Now we're a bit surprised by his insistence. We choose not to be pedantic, and skip the question of whether he means "the quantity of negative numbers is not less than zero" or "the value of any negative number is not less than zero," for clearly he means the latter. Still in disbelief, perplexed, we ask: "Are you saying, really, seriously, that the proposition

$$-1 < 0$$

is *false*?" And still smiling, he replies "*Yes*." So now we begin to wonder whether his smile is a sign of insanity. We do not want to go through the hassle of demonstrating that -1 is less than zero, but we want to be civil. So we say: "Do you agree that four is less than five?" And he replies: "Yes." So we proceed, "Well then, given that

$$4 < 5,$$

we may subtract five from both sides:

$$4 - 5 < 5 - 5,$$

so that clearly,

$$-1 < 0.$$

We have proven it because the operation of subtraction performed equally on both sides of an inequality preserves the inequality." But the guy is still smiling, and says, "The claim that equal subtractions preserve the inequality is merely a rule that you have *assumed* as if it were some sort of universal law. But actually, we may well assume instead that it is only true in *some* cases."

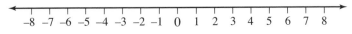

Figure 6. A number line.

By this point our patience is failing, so, frustrated, we ask: "Can you show us any specific way, any sense, in which we may properly say that −1 is not less than zero?"

"Very well, yes," replies the mathematician, "I can show you, with a simple example, not merely that −1 is not less than zero but that it can just as well be deemed to be *greater than zero.*" Now *this* sounds like a truly insane claim. Yet he insists that he will demonstrate it not even by alluding to any paradoxes about negatives, but by using reason and analogy. So again we are asked to consider the number line of figure 6. And we are asked to imagine three people positioned at the coordinate zero. Now, one of them moves four steps to the right, so we say: positive four. Another moves five to the left, so we say: negative five. And the third person does not take any steps, but remains at zero. So we're asked: "Who moved least of all?" We answer that, clearly, the person who remained at zero. Now the mathematician concludes: "Accordingly, zero is less than negative five, and likewise, thus, *negative one is not less than zero.* Moreover, here we see that negative five can be *greater* than positive four. There is, truly, nothing special about motions to the right as opposed to motions to the left. The two are entirely symmetric. The two orientations are perfectly equivalent, so it is merely an arbitrary assumption to establish that positives are greater than zero, and negatives are less."

Hum. Well. We might still question such arguments. We can ask: "Are you saying then, that

$$-5 = +5,$$

are you saying this?" To which the mathematician can reply: "That depends on the definition that we *choose* to give to the = sign. If we say that it means only equality as to quantity, then yes, the proposition is true. But if we mean equality of direction also, then *no*, since the directions are opposite, the two are not equal.

But in this case, just because the two are not equal does not mean that either one is greater than the other; they're just opposite."

We might still be annoyed at the guy's attitude. We do not want to be convinced by nonexistent mathematicians that their apparently nonsensical propositions are perhaps just as reasonable as ours. But finally, if only to end the conversation at this point, we decide to agree that, okay, there is here at least one sense in which we can say that negative numbers are not less than zero.

Of course, not everyone will let it rest at that. Someone could still argue that the demonstration is unsatisfactory because it is based completely on physical notions, and that mathematical proof does not rest on such notions. So it might seem. However, such physical notions can well be replaced by abstract principles designed to give the same results. After all, *that is precisely how many of the elementary principles of traditional mathematics originated.*

TRADITIONAL COMPLICATIONS

If we are to devise math rules by tampering with traditional principles, we might perhaps need some strong motivation. After all, despite any ambiguities in some notions and definitions, we must admit that ordinary traditional arithmetic and algebra are immensely useful and apparently unproblematic. And even if we empathize with some of the objections raised by mathematicians in the past, we might not be compelled to tinker with the well-established rules solely on the basis of such objections.

Still, should only the very best mathematicians appraise the principles of mathematics? Are only they qualified to judge the simplicity and necessity of the elementary rules, with due regard to the farthest reaches of mathematical structures? In one view, yes. But in an old view, everyone is esteemed capable of appraising critically at least the elementary principles. In this conception, mathematical elements would be simple, evident, and related to well-known common notions. Hence, for example, in his *Defense of Free-Thinking in Mathematics*, George Berkeley sought even to encourage those, who are not far gone in these Studies, to use intrepidly their own judgement, without a blind or a mean deference to the best of Mathematicians, who are no more qualified than they are, to judge of the simple apprehension, or the evidence of what is delivered in the first elements of the method; men by further and frequent use or exercise becoming only more accustomed to the symbols and rules, which doth not make either the foregoing notions more clear, or the foregoing proofs more perfect.[1]

Berkeley argued that in a free country all people are entitled to use their reason and common sense to responsibly appraise the foundations of mathematics. He hoped that impartial readers would not learn by rote, or take principles on trust of authority, but would freely strive to distinguish the evident parts of mathematics from the obscure. Accordingly, we may well proceed to examine some of the ambiguities and complications in the elements of algebra.

Recall now the difficulties introduced in the first chapter. We saw that the traditional rules on the operation of positive and negative numbers lead to various formal complications and asymmetries. We can now consider a few more examples in light of the history of algebra. Great complications stem from some apparently simple expressions and rules. We usually render such complications invisible by arbitrarily and systematically disregarding them. But let's look at them for a moment.

For example, what is the algebraic meaning of the expression $\sqrt{-1}$? The answer we give depends on how seriously we take the question. A simple answer is to say that

$$\sqrt{-1} = i.$$

But this answer hides a complication; namely, that there are two interpretations of i that are usually confused with one another. First, i often is used as a compact sign, *an abbreviation*, that stands for the expression $\sqrt{-1}$. Second, i sometimes is used to represent *a solution* of the operation "extract a square root of negative one." The two interpretations are not equivalent because i is not the only solution of $\sqrt{-1}$. Remember that $-i \times -i = -1$; therefore,

$$\sqrt{-1} = -i.$$

So it seems that $\sqrt{-1}$ has two solutions. If we care enough to apply this algebraic rule rigorously, then it quickly leads to further complications. For example, consider the addition of two complex numbers:

$$(3 + 4\sqrt{-1}) + (5 + 2\sqrt{-1}) = ?$$

If we disregard the exact meaning of $\sqrt{-1}$, we can simply write

$$(3 + 4i) + (5 + 2i) = 8 + 6i,$$

where the symbol i is simply an abbreviation. But if we take the two solutions of $\sqrt{-1}$ into account, then we should write

$$(3 \pm 4i) + (5 \pm 2i),$$

so that according to how we add or subtract the imaginary terms we then have *four* solutions to the addition of the complex numbers. And any further operation with complex numbers will multiply the possible number of solutions accordingly. Before we pursue this consequence with examples, we still have plenty of complications to worry about at the simplest level.

In particular, we allowed that $\sqrt{-1}$ has two solutions; but why stop there? What about quaternions? Remember that

$$i^2 = j^2 = k^2 = -1.$$

Therefore,

$$\sqrt{-1} = j,$$

and also

$$\sqrt{-1} = k.$$

So now it seems that $\sqrt{-1}$ has four solutions. But wait, what works for i also works for j and k, so

$$\sqrt{-1} = -j$$

and

$$\sqrt{-1} = -k.$$

So it has six solutions? But furthermore, $ijk = -1$, so

$$\sqrt{-1} = \sqrt{(ijk)}$$

and

$$\sqrt{-1} = -\sqrt{(ijk)}.$$

So $\sqrt{-1}$ has eight solutions. Or does it have more? *It does.* You figure them out! Moreover, is there anything that prevents us from inventing new solutions, such as

$$\sqrt{-1} = l,$$

and so on? Nothing, aside from our habitual satisfaction with preestablished rules.

"But wait," you may say, "this all has to be wrong; isn't it true that since $i^2 = j^2$, then $\sqrt{(i^2)} = \sqrt{(j^2)}$, so that $i = j$?" No. The imaginary values i, j, k are not equal to one another. This is a strange aspect of quaternions, yes, but we find it also in traditional algebra. For example, $i^2 = (-i)^2$, but $i \neq -i$. This all stems from the traditional arithmetical rule that, for example,

$$(5)^2 = (-5)^2,$$

while

$$5 \neq -5.$$

Yes, *all* of this, all these complications stem from the idea that $(a)^2 = (-a)^2$. They stem from the asymmetric pair of rules that "plus times plus is plus," and "minus times minus is plus."

You might still think that most of these complications pertain to quaternions rather than traditional algebra, and that quaternions need not be introduced. But the distinction is arbitrary. You *choose* what level of complexity you want to admit. A complex number is simply a quaternion, $w + ix + jy + kz$, where $y = 0$ and $z = 0$. The rules of operation are all the same as in traditional algebra.

Furthermore, let us consider now a couple of asymmetries relating to the multiplicity of solutions. Remember that $\sqrt{+1}$ has only two solutions. Meanwhile, $\sqrt{-1}$ has many. Likewise, ever since Euler's work, the expression $\log_2(4)$ has infinitely many imaginary solutions *and* one positive solution. Meanwhile, $\log_2(-4)$ has infinitely many imaginary solutions but *no* positive or negative solutions. This disparity in the quantity of solutions pertaining to *the same* operations carried out on positive and negative numbers is by itself a noteworthy asymmetry.

The multiplicity of solutions of some equations is a common feature in various branches of mathematics. For example, inverse trigonometric functions have infinitely many solutions. Whereas $\sin \alpha$ equals y, where y has only one value for a given angle α, the inverse operation, $\sin^{-1} y$ has infinitely many values. As for the roots of numbers, the multiplicity of solutions increases as radicals of indices greater than $\sqrt[2]{}$ are considered. For example, the number 1 has three cubic roots:

$$\sqrt[3]{1} = 1, \frac{-1 + \sqrt{-3}}{2}, \frac{-1 - \sqrt{-3}}{2},$$

where, as usual, we let the symbol $\sqrt{-3}$ designate only a single value. Likewise, other numbers also have three cubic roots, for example,

$$\sqrt[3]{8} = 2, -1 + \sqrt{-3}, -1 - \sqrt{-3}.$$

Likewise, all cubic equations are said to have three roots. This multiplicity of roots comes from complex numbers. Such complex roots are often devoid of meaning in ordinary practical problems. For example, to figure out the length of a side of a cube that comprises a given volume, we extract the real cubic root, but we disregard the complex roots.

To avoid multiple solutions, mathematicians and scientists have employed various conventions. One approach is to distinguish between two operations: extraction of any roots and extraction of the so-called principal root. For example, Giuseppe Peano used asterisks to distinguish the two operations. The present book, as noted at the outset, uses the radical sign $\sqrt{}$ to mean the unrestricted operation and the same sign with a vinculum $\sqrt{}$ to mean the principal root. A more common practice among writers has been to use the radical sign only for the principal root, and then just not designate any sign for the unrestricted operation. This approach, however, is problematic, especially when writers use the radical sign in one place to extract only positive square roots, and subsequently use the same sign, inconsistently, to extract also complex cube roots.

Another way to avoid multiple solutions involves the concept of function. Some writers defined function as a rule that assigns elements of one set of numbers to elements of another set. There could then be two square root "functions": one that gives pairs of numbers along the positive and negative range, and another that gives only non-negative results. In contradistinction, many other writers have defined *function* as a rule that assigns to each element of one set only *one* element of another set. For example, Jean Dieudonné advocated a definition of mapping that allows exclusively single-valued functions, "despite many books to the contrary."[2] Thus he opposed the practice of mathematicians who allow multivalued functions as in the old definition of the radical operation. Dieudonné's concern illustrates the point that it is often inconvenient to employ multivalued operations.

In textbooks on calculus, especially, functions are usually defined to assign only one value $f(x)$ for each x. In particular, the ambiguity of double solutions is often avoided by defining the square root operation as

$$\sqrt[2]{x^2} = |x|,$$

where x is not an imaginary or complex number.[3] Here, the square root is a function that is restricted in two ways. The domain of values that x can have is delimited to include only positive and negative numbers and zero. And the range of values that can result are restricted to only the non-negative, by using the absolute value operation. By this convention, writers hence allow the root function to yield only one value for each x. Thus, students avoid multiple roots, negative roots, and complex roots.

Despite such common attempts to delimit the use of the radical sign, and to restrict the domain and range of some functions, ambiguities continue to lurk regarding multiple values. The preferential treatment granted to positive values does not solve such ambiguities, it merely sets them aside. Yet what happens when we allow algebra its multiple solutions?

Some mathematicians have claimed that problems then emerge even in basic operations. For example, consider again the addition of square radicals. Dieudonné argued that if indeed we allow the function $\sqrt{z}$ to have two values for each z, then "The penalty for this indecent and silly behavior is immediate: it is impossible to perform even the simplest algebraic operations with any reasonable confidence: for instance $2\sqrt{z} = \sqrt{z} + \sqrt{z}$ is certainly *not true*, for the definition of $\sqrt{z}$ we are compelled to attribute *two* distinct values to the left hand side, and three distinct values to the right hand side!"[4]

His point may be clarified. He assumed that if we let $\sqrt{z} = \pm r$, then

$$2\sqrt{z} = +2r, -2r,$$

that is, it has two values. Meanwhile, he also assumed that the expression $\sqrt{z} + \sqrt{z}$ could be solved in four different ways: $r + r$, $r + -r$, $-r + r$, and $-r + -r$. It would thus produce three results:

$$\sqrt{z} + \sqrt{z} = +2r, 0, -2r.$$

Therefore the algebraic proposition $2\sqrt{z} = \sqrt{z} + \sqrt{z}$ would not be generally valid, just because $2\sqrt{z}$ does not equal 0.

But contrary to Dieudonné's intent, his argument is inadequate. To claim that $\sqrt{z} + \sqrt{z}$ has *three* different values, we have to make an independent and dispensable assumption: that two z's in a single proposition can have distinct values *at the same time*. Without this alien requirement, $\sqrt{z} + \sqrt{z}$ has only two possible forms:

$$\sqrt{z} + \sqrt{z} = (+\sqrt{z}) + (+\sqrt{z}) = +2\sqrt{z}$$

and

$$\sqrt{z} + \sqrt{z} = (-\sqrt{z}) + (-\sqrt{z}) = -2\sqrt{z},$$

such that now, $\sqrt{z} + \sqrt{z}$ certainly equals $2\sqrt{z}$, because

$$(\pm\sqrt{z}) + (\pm\sqrt{z}) = \pm2\sqrt{z}.$$

Thus there is a straightforward and consistent way to add multiple radicals. Still, the point is that there exist confusions in how to deal with operations that entail multiple solutions.

So what now? Suppose you are writing a basic textbook on algebra. You reach the point where you have to set forth the rule for adding radicals. It would seem now that you have at least three options on how to proceed. (1) Avoid the subject of multiple solutions by defining the radical operation to extract only one root. (2) Admit multiple roots and employ an addition rule that adds all four possible additions of double signs. (3) Admit multiple roots but restrict addition such that it allows only two additions of double signs. Each approach involves complications. The first simply disguises or postpones the ambiguity. The second creates a multiplicity of solutions that then undermines the general validity of some equations such as $2\sqrt{z} = \sqrt{z} + \sqrt{z}$. The third seems elegant, but entails a complication that we will see below.

To further illustrate ambiguities rooted in the rules of signs, consider next the operation of multiplication. Consider the algebraic rule

$$\sqrt{a} \times \sqrt{b} = \sqrt{(ab)}.$$

In 1770, Euler stated that this rule is valid regardless of whether a and b are positive or negative.[5] Nowadays, however, historians and mathematicians routinely note that this rule is actually not valid for negative numbers. For example, given two negative numbers

$$\sqrt{-4} \times \sqrt{-9} \stackrel{?}{\Leftrightarrow} \sqrt{(-4 \times -9)}$$

$$2i \times 3i \stackrel{?}{\Leftrightarrow} \sqrt{36}$$

$$-6 \neq 6.$$

Thus, if a and b are negative,

$$\sqrt{a} \times \sqrt{b} \neq \sqrt{(ab)}.$$

So, writers claim, Euler was simply confused and mistaken.[6]

But the question is not as simple as it appears. Notice that by writing 6 as the square root of 36 we neglected the negative root −6. Likewise, in solving the left side of the equation, we employed the interpretation of i as an *abbreviation* of $\sqrt{-1}$, neglecting to note that it actually has two solutions. Indeed, as Euler himself indicated, *every* square radical has two solutions (except for the number 0).[7] Taking these points into account, we should properly write

$$\sqrt{-4} \times \sqrt{-9} \overset{?}{\Leftrightarrow} \sqrt{(-4 \times -9)}$$

$$(\pm 2i) \times (\pm 3i) \overset{?}{\Leftrightarrow} \pm 6,$$

where $\sqrt{-4}$ and $\sqrt{-9}$ each have two solutions. Now notice that there are four ways to multiply the numbers $+2i$, $-2i$, $+3i$, $-3i$:

$$(+2i) \times (+3i) = -6$$

$$(+2i) \times (-3i) = +6$$

$$(-2i) \times (-3i) = -6$$

$$(-2i) \times (+3i) = +6.$$

And these results are summarized by the expression

$$(\pm 2i) \times (\pm 3i) = (\pm 6),$$

and therefore, substituting into the previous equation,

$$(\pm 6) \overset{?}{\Leftrightarrow} \pm 6$$

$$\pm 6 = \pm 6.$$

So, finally, the equation

$$\sqrt{a} \times \sqrt{b} = \sqrt{(ab)}$$

is actually true for negative as well as positive numbers. So Euler was right. This result should be surprising, considering that we are discussing a basic rule of ordinary algebra. Have mathematicians, then, been confused about the matter for over two hundred years? This idea may seem so ludicrous that you may wonder whether there is some mistake in the present analysis.

However, Euler's rule is actually so fundamental that it underlies even the common practice of loosely writing, say,

$$\sqrt{-4} = 2i.$$

Because, to extract the number 2 from under the radical sign we must require that

$$\sqrt{-4} = \sqrt{(4 \times -1)} = \sqrt{4}\sqrt{-1},$$

which is nothing other than the equation in question:

$$\sqrt{(ab)} = \sqrt{(a \times b)} = \sqrt{a}\sqrt{b}.$$

Indeed, it was on the basis of this very rule that Euler justified the extraction of a positive factor from the radical of a negative number.[8] Now, if we carry out the established rules rigorously, we finally have to write

$$\sqrt{4}\sqrt{-1} = \pm 2\sqrt{-1}.$$

You might now notice that the rule in question may be employed in the previous demonstration:

$$\sqrt{(4 \times -1)} \times \sqrt{(9 \times -1)} \overset{?}{\Leftrightarrow} \sqrt{(36)}$$

$$\sqrt{4}\sqrt{-1} \times \sqrt{9}\sqrt{-1} \overset{?}{\Leftrightarrow} \pm 6$$

$$(\pm 2\sqrt{-1}) \times (\pm 3\sqrt{-1}) \overset{?}{\Leftrightarrow} \pm 6.$$

But we avoided this procedure because we sought to justify the very rule, which we did by appealing instead to an independent rule: that extracting square roots yields two solutions. Accordingly, Euler emphasized that "by the nature of the radical sign, $\sqrt{-1}$ encloses essentially the sign + as well as the sign –," that is, two solutions. Euler emphasized that there is *always* an even number of imaginary roots, and *never* an odd number.[9]

Mathematicians soon proposed a rule different from the one

provided by Euler; as though he had simply committed a mistake.[10] They argued that for negative numbers

$$\sqrt{a} \times \sqrt{b} = -\sqrt{(ab)}.$$

Euler's rule was rejected on the basis of the notion that

$$\sqrt{-1} \times \sqrt{-1} = \sqrt{(-1)^2} = -1.$$

But, strictly speaking, this notion is dispensable; it is an independent assumption.

In the early 1800s, many mathematicians regarded the equation $\sqrt{-1} \times \sqrt{-1} = -1$ as necessarily and exactly true. But following the rise of symbolical algebra, in the works of George Peacock, Augustus De Morgan, and others, suchlike equations came to be increasingly regarded as arbitrary stipulations. For example, Peacock claimed that unlike the results of arithmetical algebra, the results of symbolical algebra are not necessary, they are not unavoidable consequences of our numerical conceptions. He argued that the forms or results of symbolical algebra that do not mimic those of arithmetical algebra are hence "conventional."[11] Likewise, later in the 1800s, some mathematicians characterized the requirement that $\sqrt{-1} \times \sqrt{-1} = -1$ as a useful "convention" or "supposition."[12]

One way to replace this convention is to take into account the rule that every square root has two solutions. Thus, to take a simpler example, we might likewise imagine that

$$\sqrt{1} \times \sqrt{1} = 1,$$

but if we actually solve the square roots, we obtain

$$\sqrt{1} \times \sqrt{1} = \pm 1 \times \pm 1$$

$$\sqrt{1} \times \sqrt{1} = \begin{bmatrix} 1 \times 1 = 1 \\ -1 \times 1 = -1 \\ 1 \times -1 = -1 \\ -1 \times -1 = 1 \end{bmatrix}$$

$$\sqrt{1} \times \sqrt{1} = \pm 1.$$

Meanwhile,

$$\sqrt{(1^2)} = \sqrt{1} = \pm 1.$$

Therefore

$$\sqrt{1} \times \sqrt{1} = \sqrt{(1^2)} = \pm 1.$$

And by a parallel argument

$$\sqrt{-1} \times \sqrt{-1} = \pm i \times \pm i$$

$$\sqrt{-1} \times \sqrt{-1} = \begin{bmatrix} i \times i = -1 \\ -i \times i = +1 \\ i \times -i = +1 \\ -i \times -i = -1 \end{bmatrix}$$

$$\sqrt{-1} \times \sqrt{-1} = \pm 1,$$

while

$$\sqrt{(-1^2)} = \sqrt{1} = \pm 1.$$

Therefore

$$\sqrt{-1} \times \sqrt{-1} = \sqrt{(-1^2)} = \pm 1.$$

By contrast, equations such as

$$\sqrt{-1} \times \sqrt{-1} = \sqrt{(-1^2)} = -1$$

and

$$\sqrt{-1} \times \sqrt{-1} = (\sqrt{-1})^2 = -1$$

are true only in a special case, that is, only *if* the sign $\sqrt{}$ attached to a number is interpreted to represent *a single solution*, that is, a single number, rather than an operation to be performed on the number. A single number multiplied by a single number yields a single result. But the extraction of square roots of a number yields two results (or more, as we know).

Accordingly, the mathematician and logician Gottlob Frege cautioned: "Nothing prevents us from using the concept 'square root of −1'; but we are not entitled to put the definite article in front of it without more ado and take the expression 'the square root of -1' as having a sense."[13]

Consider now some of the subtleties of dealing with double signs. Suppose we establish, as above, that

$$\sqrt{1} \times \sqrt{1} = \pm 1 \times \pm 1 = \pm 1.$$

Compare this multiplication rule with the addition rule that we considered previously:

$$\sqrt{1} + \sqrt{1} = \pm 1 + \pm 1 = \pm 2.$$

Both of these rules seem to harmonize in that they produce double signs from double signs. However, there is a significant difference between them. The multiplication rule produces the double sign by carrying out a fourfold multiplication of signs. But the addition rule produces a double sign by carrying out only a twofold addition of signs. It presupposes that we *deny* that identical terms in one equation can have different values simultaneously. Therefore, these two rules seem to be incompatible. So it seems we have to make a choice.

If we accept the fourfold multiplication rule, we may also adopt the fourfold addition rule, such that

$$\sqrt{1} + \sqrt{1} = \pm 1 + \pm 1 = +2, 0, -2, 0,$$

where we again have three distinct results. Is this inconceivable? No. But is it "indecent and silly"? Is it unacceptable?

As we already saw, if we accept this fourfold addition rule we then abandon the general validity of some usual algebraic propositions, such as $2\sqrt{z} = \sqrt{z} + \sqrt{z}$. Well, suppose we do. We should then not assume that the usual rules of arithmetical algebra are all valid for the unrestricted algebra of signs. We would have to explore what algebraic rules are generally valid, and which ones are valid only in special cases. It may seem unusual. But this is just how ordinary algebra works. The equation $x^2 = x$ may seem ridiculous if we posit it as a general law of numbers. But it is valid in the special cases where x equals 0 or 1. Likewise, if algebra admits the fourfold addition of double signs then the equation $2\sqrt{z} = \sqrt{z} + \sqrt{z}$ acquires only a limited validity.

Now suppose that, instead, we deny the fourfold addition and accept the double addition rule of signs. It seems reasonable to expect that then we would have to reject the general validity of the rule $\sqrt{a} \times \sqrt{b} = \sqrt{(ab)}$. It would still be valid in some cases, such as when only positive numbers are concerned.

But there are other ways to proceed as well. Can we possibly accept both the fourfold multiplication rule and the twofold addition rule? Yes. One has only to realize that what works for addition need not work in the same way for multiplication. After all, they are distinct operations. For example, multiplication has the numerical property that it is distributive:

$$a \times (b + c) = (a \times b) + (a \times c),$$

whereas if we write

$$a + (b \times c) = (a + b) \times (a + c),$$

we find that, generally, this proposition is numerically false. Addition is not distributive. Likewise, it is conceivable that we might admit the equations $\pm1 \times \pm1 = \pm1$ and $\pm1 + \pm1 = \pm2$, simultaneously, if only we are ready to allow that multiplication and addition do not summarize the handling of double signs in the same way.

In any case, we have to make some choices. The various alternatives each entail diverse simplifications and complications. Which rules we choose to assert may depend on which seem to be more convenient. It also depends on which complications seem more familiar or palatable than others.

Moving along, consider yet another ambiguity. Briefly, let's further consider the expression

$$\sqrt{(a^2)} = a.$$

This expression was asserted as a rule by various algebraists in the old days, such as Etienne Bézout (one of the educators disliked by Henri Beyle) and S. F. Lacroix.[14] This equation is still used routinely though tacitly by many mathematicians and physicists to conveniently assume that extraction of square roots and raising numbers to the second power are exactly inverse operations. They might not formulate it expressly, but they use it whenever they need to reduce a squared variable to itself. But strictly speaking, this rule can lead to contradictions with a fundamental rule of arithmetic: that equal operations performed on both sides of an equation always preserve the equality. For example, take the equation

$$(+2)^2 = (-2)^2,$$

and, by applying the traditional rule above to both sides of it,

$$\sqrt{((+2)^2)} = \sqrt{((-2)^2)},$$

we obtain

$$+2 = -2,$$

a clear contradiction. Hence it might be preferable to expect instead, in accordance with Euler's product rule, that

$$\sqrt{a} \times \sqrt{a} = \sqrt{(aa)} = \pm a,$$

so that

$$\sqrt{((+2)^2)} = \sqrt{((-2)^2)}$$

$$\sqrt{(+4)} = \sqrt{(+4)}$$

$$\pm 2 = \pm 2.$$

This procedure is completely symmetric. And so the fundamental rule now holds: that equal operations carried out on both sides of an equation preserve the equality.

Whatever may be the significance of the general validity of the equations $\sqrt{z} + \sqrt{z} = 2\sqrt{z}$, $\sqrt{a} \times \sqrt{b} = \sqrt{(ab)}$, and $\sqrt{a^2} = \pm a$ for traditional algebra, the reason why they have been discussed here is mainly to illustrate how subtleties in the traditional rules of signs can yield profound ambiguities. Asymmetries yield complications, which can admit ambiguities, which sometimes entail mistakes. Thus we should try to elucidate the exact consequences of our rules more carefully and explicitly. Or we might also choose to develop new rules that do not lead so easily to ambiguities.

Return now to the question of ascribing new meaning to traditional concepts. We said that we can invent, for example, new solutions for the expression $\sqrt{-1}$. Actually, we need not even limit ourselves to purely mathematical concepts. In this regard, Frege argued that we might freely establish new definitions for $\sqrt{-1}$ by letting -1 have new "square roots," such as a unit of electricity, a certain surface area, or even the Moon:

> We take some object, let us say the Moon, and proceed by defini-
> tion: Let the Moon multiplied by itself be -1. This gives us a
> square root of -1 in the shape of the Moon. There seems to be
> nothing wrong with this definition, since meaning hitherto as-
> signed to multiplication says nothing as to the sense of a product
> such as the Moon into the Moon, so that as we now come to ex-
> tend its meaning we can make it, for the Moon, whatever we
> choose.[15]

Irrespective of how useless or silly this particular suggestion may seem, the important point is to illustrate that we have some free-dom to develop concepts as we see fit. Frege continued: "That we are able, apparently, to create in this way as many square roots of -1 as we please, is not so astonishing when we reflect that the meaning of the square root of -1 is not something which was already unalterably fixed before we made these choices, but is decided for the first time by and along with them." By making the Moon a square root of -1, we have extended the meaning of the concepts of root extraction and multiplication. In a sense we have "changed" these concepts, but *still*, we expect the usual re-sults to apply when we carry out operations such as $\sqrt{4}$ or -2×-2. We expect, essentially, that by extending the meaning of the expression $\sqrt{-1}$ we do not generate any contradictions or incon-sistencies with our preestablished rules.

This is roughly what happened with quaternions: the new square roots of -1 did not contradict the previous rules on the op-eration of "traditional" imaginary numbers considered by them-selves. One could still then say that i times i is equal to -1. But the general algebraic equation, $i \times a = a \times i$, was no longer true for all numbers a, since now there were new numbers j and k that did not fulfill this proposition.

But this is only one way of modifying the meaning of familiar concepts. This process of modification by extension, that is, by defining new additional meanings, has many uses, but we must note that it leads to new complications, since we are essentially adding new complexity while keeping the preestablished definitions. Another way to proceed is to reduce the complexity by restricting the meaning of terms. Maybe we don't want the Moon to be a square root of anything.

Furthermore, we may perhaps also devise alternative definitions of traditional concepts that deviate considerably from the traditional rules. Remember: "it must be observed that we have a power over the definitions," in the words of the Cambridge mathematician Isaac Todhunter, "It is therefore in our power to define them as we please provided we always adhere to our definition."[16]

CAN MINUS TIMES MINUS BE MINUS?

As we have seen, the general application of ordinary algebraic rules entails various sorts of mathematical elements, such as imaginary numbers and quaternions. Accordingly, it also entails variations on the effects and validity of preestablished arithmetical operations and notions.

As suggested earlier, we might be able to dispose of all such complications if we could somehow change the rules that lead to the "existence" of imaginary expressions. And there is but one rule that leads to the introduction of imaginaries, namely, the extraction of roots of negative numbers. So let's just change that rule, to see what happens. We will then get some basic impressions of what an algebra based on unusual numerical rules would be like. Likewise, recall that for centuries mathematicians and teachers complained about double or multiple solutions of equations. Accordingly, let us experiment with rules designed to entail only a single solution for every equation. For the sake of the argument, we will thus try to construct an artificial rudimentary arithmetic and a corresponding new algebra.

We should specify what we mean by the words "arithmetic" and "algebra." Owing to the old ambiguities relating to negative numbers and more, mathematicians traditionally restrict arithmetic as pertaining only to certain sorts of numbers, usually positive whole numbers and zero, and only to certain operations, such as addition, subtraction, multiplication, and division. As for algebra, mathematicians have defined it in very many distinct and discordant ways. To avoid such ambiguities and arbitrary restrictions, we may adopt more convenient and simple definitions. "Arithmetic" here means the study of the relations among numbers and operations. "Algebra" means the study of such numerical relations and operations where at least some numbers are not specified directly but are represented by letters.

The task of devising principles that are best suited for describing things of experience is postponed for the following chapter. For now, we will modify the traditional rules of signed numbers, that is, we will invent artificial rules, chiefly as a basic exercise demonstrating the flexibility or plasticity of the elements of mathematics. Hence we will, for now, view algebra essentially as a system of symbols. Nevertheless, we do not want these symbols to be completely devoid of meaning. Therefore, even though we will develop fanciful algebraic rules, we will aim to make those rules consonant with elementary arithmetical and geometrical notions.

To begin, instead of traditional rules such as

$$\sqrt{-1} = i,$$

let us stipulate that now

$$\sqrt{-1} = -1.$$

Accordingly, let us change the traditional rule that

$$\sqrt{+1} = \pm 1,$$

with its ambiguity about the double sign, that is, of the two possible solutions, to establish instead the simple rule that

$$\sqrt{+1} = +1.$$

If we can do this, without introducing inconsistencies and contradictions into our imaginary algebra, we will no longer need to conjure imaginary numbers for the solution of some problems. Thus we might dispense with complications embodied by the theory of complex numbers. By formulating fictitious rules, and tracing some of their consequences, we will find whether they are indeed plausible, at least to some degree. Afterward, we will seek algebraic generalizations.

In accord with our makeshift rules on the extraction of roots, we can also stipulate the following rules:

$$(+1)^2 = +1, \quad (+2)^2 = +4, \quad \text{etc.,}$$

and

$$(-1)^2 = -1, \quad (-2)^2 = -4, \quad \text{etc.,}$$

where the parentheses have been included merely to indicate clearly that the operation of squaring is to be performed on these signed numbers. These operations are simply the reverse of the corresponding square root extractions. Note that here we need not assume that the exponent 2 is positive or negative. It is merely a symbol to indicate the operation of taking the second power of a number. Thus its effect is now symmetric on positive and negative numbers. To be sure, we might as well introduce positive and negative exponents also. We may establish that

$$(+1)^{+2} = (-1)^{+2} = +1, \quad \text{etc.,}$$

and, likewise, that

$$(-1)^{-2} = (+1)^{-2} = -1, \quad \text{etc.,}$$

and we may call these results the "positive and negative powers" of any number.

Notice that thus far every improvised rule on how to operate tentatively with positive and negative numbers involves results that are entirely symmetric. The artificial arithmetic we are devising does not involve some of the basic asymmetries inherent in the traditional formalism. For example, it does not involve asymmetric pairs of propositions such as

$$(+2)^{+2} = +4, \quad (-2)^{-2} = +\tfrac{1}{4}.$$

Now of course, if we wish to replicate results such as $+\tfrac{1}{4}$, we need only define further artificial operations that yield them. But instead of proceeding down that path, let's turn to the more difficult and important question of establishing rules for the combination of positives and negatives. We might fear that in the attempt to establish such combinations, in accord with the fabricated rules, we will encounter insurmountable difficulties.

Naturally, we should expect that underlying the rules about exponents and square roots will be the rules of multiplication. But before discussing that, let's take a moment to consider the operations of addition and subtraction.

Recall that the traditional rules on the addition of positives and negatives lead to results that are quite symmetric. For example,

$$^+5 + {}^+3 = {}^+8$$

$$^-5 + {}^-3 = {}^-8,$$

and, in the combination of signs,

$$^+5 + {}^-8 = {}^-3$$

$$^-5 + {}^+8 = {}^+3.$$

The only asymmetry here is merely apparent: it is the use of the same symbol to designate addition as well as positives. The two concepts, however, are distinct, and thus if we replace, say, the traditional symbol of addition with a new one, any apparent asymmetries disappear. For example, we could write

$$^+5 \wedge {}^-8 = {}^-3$$

$$^-5 \wedge {}^+8 = {}^+3,$$

and thus see that the two expressions are fully symmetric. We will not change the traditional symbols, however, just because we'd like to retain the familiarity of the usual expressions. Hence we shall continue to uphold the old hope that the rather unfortunate use of the same symbol for distinct concepts will not lead us astray. Regardless, the important point is that the traditional rules on the addition of signed numbers lead to symmetric results. Therefore, we will incorporate all such traditional rules into our artificial arithmetic.

Likewise, the traditional operation of subtraction also leads to symmetric results. So we will adopt its usual rules as well. Nevertheless, we should note that we need not assume the traditional association of subtraction with multiplication of -1. Specifically, we need not adopt the traditional rule expressed in equations such as

$$^+5 - {}^-8 = {}^+5 + (^-1) \times (^-8).$$

We will adopt equations such as

$$^+5 - {}^-8 = {}^+5 + {}^+8,$$

but we will not interpret this "change of signs" as implying the action of any implicit multiplication. Therefore, we will understand the artificial operation of subtraction as follows. In any expressions such as

$$- {}^-8$$

we will interpret the sign of subtraction as requiring that we change the sign of the number following it. And, in any expressions such as

$$^+5 - {}^-8$$

we will interpret the sign of subtraction as requiring that we *change* the sign of the number following it and *then combine* that number with the preceding number. Accordingly, we also can read any expressions such as

$$+ {}^-8$$

as requiring that we keep the sign of the number following it; and we may interpret any expressions such as

$$^+5 + {}^-8$$

as meaning that we *keep* the sign of the number following it and *then combine* that number with the preceding number. Again, these operations, their symbolism, and interpretations are symmetric.

By this point you might realize that although we said that we would develop a system that is essentially symbolic, the operations defined thus far can be understood in various ordinary ways. You might, for example, entertain ideas about motions in opposite directions as you consider the improvised rules for the combination of positives and negatives. Indeed, as in the early attempts at formulating "purely" symbolic algebras, we are cheating discreetly by borrowing principles from experience and pretending that they are purely abstract and independent of experience. Thus we now adopt the common empirical idea that two plus two is four, as well as the not obvious notion that minus five and plus five are both equally greater than zero.

Having admitted such empirical grounds, the mathematics being developed is, nevertheless, essentially symbolic. Because, we define its fundamental concepts irrespective of any specific context of meaning or connotations. If a set of rules is thus treated abstractly, without restrictions to any specific interpretation, then it constitutes a symbolic mathematics. The task of providing meaningful interpretations of any specific operation is then left open for anyone who wishes to apply that mathematics to a particular sort of problem. Augustus De Morgan provided a nice illustration of how a symbolic relation of combination can be ascribed a variety of valid interpretations:

> Given symbols M, N, +, and one sole relation of combination, namely, that $M + N$ is the same result (be it of what kind soever) as $N + M$. Here is a symbolic calculus: how can it be made a significant one? In the following ways, among others. 1. M and N may be *magnitudes*, + the sign of addition of the second to the first. 2. M and N may be *numbers*, and + the sign of multiplying the first by the second. 3. M and N may be *lines*, and + a direction to make a rectangle with the antecedent for a base, and the consequent for an altitude. 4. M and N may be *men*, and + the assertion that the antecedent is the brother of the consequent. 5. M and N may be nations, and + the sign of the consequent having fought a battle with the antecedent: and so on.[17]

Thus a symbolic algebra can be susceptible to a diversity of meaningful interpretations. Yet we define its rules without ascribing fundamental or general importance to any particular empirical analogy.

Consider now the operation of multiplication. The traditional rules for the multiplication of positive and negative numbers can be described as follows:

$$+ \times + = +$$

$$+ \times - = -$$

$$- \times + = -$$

$$- \times - = +.$$

(By the way, there is nothing innocent about these operations as they here stand alone, without numbers; in math and in physics we sometimes operate on operations themselves rather than on objects such as numbers.) Oftentimes these four rules have been summarized by the two expressions "like signs" multiplied yield positive, and "unlike signs" multiplied yield negative. Clearly, the last of these rules, "minus times minus is plus," at least, is incompatible with the algebra we are trying to devise. That is, if we want to retain the basic usual relationship between the operation of multiplication and that of exponents, we should now expect that, for example,

$$-2 \times -2 = (-2)^2 = -4.$$

So we would now suppose that "minus times minus is minus." In the meantime, nevertheless, we may retain the usual rule that "plus times plus is plus," since we're happy with propositions such as

$$+2 \times +2 = (+2)^2 = +4.$$

Hence we have the rules for the multiplication of "like signs."

Now we must decide what rules to employ in the multiplication of "unlike signs." How should we define the new results of the operations "$+ \times -$" and "$- \times +$"? Returning to the traditional rules for the multiplication of signed numbers, we can appreciate a nice symmetry in them: that given the four rules concerning the two signs, two results are of one sign and two are of the other sign. Wanting to adopt a similar scheme, we can contrive the following four:

$$+ \times + = +$$

$$+ \times - = +$$

$$- \times + = -$$

$$- \times - = -.$$

Here, the multiplication of "unlike signs" yields the sign of the first factor. The results of these redefined multiplication rules can now be summarized in a single statement: when multiplying two numbers, the result has the same sign as the first.

Before proceeding to analyze our fictitious multiplication rules, we might pause to note several other possibilities. Instead of the rules just chosen, and in view of slightly different considerations, we could have stipulated that

$$+ \times + = +$$

$$+ \times - = -$$

$$- \times + = +$$

$$- \times - = -.$$

Here, the sign of the second factor is the sign of the result. In this case we still retain some usual notions, such as the rule that "plus times plus is plus." Still, we might otherwise establish rules even more peculiar, by defining, for instance, that

$$+ \times + = -$$

$$+ \times - = -$$

$$- \times + = +$$

$$- \times - = +.$$

Or we might simply reverse the results of the traditional scheme, by establishing that

$$+ \times + = -$$

$$+ \times - = +$$

$$- \times + = +$$

$$- \times - = -.$$

In this case, "like signs" multiplied yield negative, "unlike signs" yield positive.

Many other seemingly strange alternative rules can be considered, especially if we abandon the requirement that the results be two of each sign. For example, Girolamo Cardano considered multiplication rules in which any number multiplied by a negative produces a negative, while only the multiplication of two positives produces a positive.[18] And Thomas Harriot contemplated the idea that all products are positive except where two negatives are multiplied.[19] In addition to such asymmetric alternative rules, we can establish even stranger rules by letting the results have entirely new signs, instead of just positive and negative. But having noted such alternative possibilities, we need not complicate the present inquiry by pursuing them. So, instead, let's continue to analyze the multiplication rules wherein the sign of the result is the sign of the first factor.

The fictitious multiplication rules entail advantages as well as disadvantages relative to the traditional rules. By this point, two of the major advantages should be obvious. First, the fictitious rules introduce certain symmetries regarding the results obtained when operating with negatives and positives. Second, in so doing, the old need for introducing imaginary numbers is entirely eliminated along with the complications of having to deal with complex number theory. But let us now consider some immediate disadvantages.

One disadvantage is that the traditional "commutative law" for the multiplication of signed numbers is not generally valid in our artificial algebra. Instead of expecting, as usual, that

$$a \times b = b \times a,$$

as for example, that

$$^+5 \times {}^-3 = {}^-3 \times {}^+5,$$

we now have

$$a \times b \neq b \times a.$$

According to the contrived rules,

$$^+5 \times {}^-3 = {}^+15$$

and

$$^-3 \times {}^+5 = {}^-15,$$

so that generally, if a and b are numbers with opposite signs, then

$$(a \times b) = -(b \times a).$$

This complicates the algebra that can result from the contrived rules; and considerably so, since the old rule

$$a \times b = b \times a$$

is still valid in some cases, that is, whenever a and b are both of the same sign.

But abandoning the commutative "law" of multiplication need not be viewed as a tragedy. Historically, it was this very abandonment that enabled Hamilton to devise his theory of quaternions, and it was this breakthrough that enabled mathematicians to realize that they could hence devise many new algebras, thus enriching the scope of mathematics. Moreover, Hamilton's noncommutative algebra proved to have many useful applications in classical as well as modern physics. Furthermore, vector algebra, which stemmed from the theory of quaternions, has been extraordinarily useful in physical applications. Vector theory likewise involved the abandonment of the commutative property for the "multiplication" of vectors. And there are other examples. Nevertheless, if we view our makeshift mathematics purely as a symbolic system, we may indeed regard the restricted commutative character of its multiplication as a disadvantage.

So there is at least one sense in which our contrived rule for the multiplication of signed numbers entails complications. How do other consequences of our rule compare with other properties of traditional multiplication? Another key property is the so-called "distributive" property. In traditional mathematics the relation

$$c(a + b) = ca + cb,$$

is said to be true for all numbers, whatever their sign. Is this same distributive relation valid in our makeshift mathematics? Consider an example, solved according to the contrived multiplication rule:

$$(-5)(-2 + 3) \overset{?}{\Leftrightarrow} (-5)(-2) + (-5)(3)$$

$$(-5)(1) \overset{?}{\Leftrightarrow} (-10) + (-15)$$

$$-5 \neq -25.$$

Thus the distributive algebraic equation above is not valid in our artificial math. So this *seems* to be another complication. But wait, this is not really the case. Since the artificial multiplication is not commutative we might now consider a different distributive property:

$$(a + b)c = ac + bc.$$

To see if it is valid, we may try any arithmetical example, as before:

$$(-2 + 3)(-5) \overset{?}{\Leftrightarrow} (-2)(-5) + (3)(-5)$$

$$(1)(-5) \overset{?}{\Leftrightarrow} (-10) + (15)$$

$$5 = 5.$$

It works! Therefore, our fictitious multiplication does entail a simple distributive property.

Finally, we may ask whether this multiplication has the "associative" property of traditional multiplication. The associative property states that for any numbers,

$$(ab)c = a(bc).$$

As before, we may test its applicability to our improvised rules by analyzing specific numerical examples, such as

$$(-2 \times 3)(-5) \overset{?}{\Leftrightarrow} (-2)(3 \times -5)$$

$$(-6)(-5) \overset{?}{\Leftrightarrow} (-2)(15)$$

$$-30 = -30.$$

Thus the artificial multiplication does involve an associative property.

We see now how our imaginary mathematics has both similarities and differences compared to traditional mathematics. But you may notice that until this point, most of the modifications to the traditional rules seem to affect only the signs of the results of operations and not the quantitative values of those results. For example, we established that

$$+3 \times -5 = +15 \text{ instead of } -15.$$

But the improvised rules change only signs only at the simplest level. By combining operations, we quickly obtain results that differ quantitatively from those obtained by traditional mathematics. For example, consider the usual solution for the following problem:

$$\sqrt{2 + \sqrt{-121}} = \sqrt{2 + \sqrt{121 \times -1}} = \sqrt{2 + \sqrt{121}\sqrt{-1}} = \sqrt{2 + 11i}.$$

By contrast, by applying our make-believe rules, we obtain

$$\sqrt{2 + \sqrt{-121}} = \sqrt{2 - 11} = \sqrt{-9} = -3.$$

The putative result differs completely from the other. It is also simpler. This suggests that very much of the complexity of traditional mathematics stems from the asymmetries in the basic rules of signs.

Such complexity is far greater than the preceding example even suggests. To further illustrate it, we may compare a few more results obtained by traditional means, with results that follow from our manufactured rules. Consider the following expressions:

$$\sqrt{+1}^{\sqrt{+1}}, \sqrt{+1}^{\sqrt{-1}}, \sqrt{-1}^{\sqrt{+1}}, \sqrt{-1}^{\sqrt{-1}}.$$

Symbolically, they are all very simple and quite symmetric. But note the extraordinary complexity that these expressions entail when solved by traditional means:

$$\sqrt{+1}^{\sqrt{+1}} = \pm 1$$

$$\sqrt{+1}^{\sqrt{-1}} = \text{infinitely many values, including,}$$
$$\text{for example, } 0.001867443 \ldots$$

$$\sqrt{-1}^{\sqrt{+1}} = \pm i, \text{ etc.}$$

$$\sqrt{-1}^{\sqrt{-1}} = \text{infinitely many values, including,}$$
$$\text{for example, } 0.207879576. \ldots$$

It is beyond the scope of this book to explain how any of the infinitely many irrational results are obtained, or how mathematicians justify them. They are included here solely to compare them with the results that you can obtain easily and immediately by applying the artificial rules:

$$\sqrt{+1}^{\sqrt{+1}} = +1$$

$$\sqrt{+1}^{\sqrt{-1}} = -1$$

$$\sqrt{-1}^{\sqrt{+1}} = +1$$

$$\sqrt{-1}^{\sqrt{-1}} = -1.$$

These results are perfectly symmetric. If we truly value simplicity in mathematics, we must agree that the make-believe rules lead to an elegant mathematics.

Mathematicians indeed often talk highly about the value of simplicity, but at the same time they also accept complexity. Thus, for example, they are fond of the number 0.207879576 . . . because although it is complicated it is useful in many analyses involving other irrational numbers such as π. So it might perhaps seem that if we establish a new value for the expression $\sqrt{-1}^{\sqrt{-1}}$ then we throw away a lot of useful mathematics. But this is not necessarily true.

We can well establish traditional results of one sort or another by new means. A simple example suffices. Consider the Pythagorean theorem (figure 7). Given a rectangular triangle, the squares

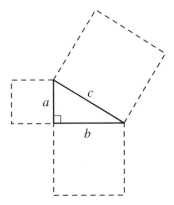

Figure 7. A diagram of the Pythagorean theorem.

constructed on its smaller sides add up to the square constructed on its largest side. Algebraically, this relation among the sides of the triangle can be expressed by

$$a^2 + b^2 = c^2.$$

Ordinarily, this equation is true regardless of the signs of the numbers a, b, c. However, consider what happens when we apply our contrived rules directly:

$$(3)^2 + (-4)^2 \overset{?}{\Leftrightarrow} (5)^2$$

$$(9) + (-16) \overset{?}{\Leftrightarrow} (25)$$

$$-7 \neq 25.$$

It seems that the Pythagorean theorem is not generally valid in regard to the fabricated rules. How can such a fundamental and ancient and true principle of mathematics not be valid? "Aha!"—you might now say, "the equation gives an *incorrect* result! Finally we see what nonsense really follows from changing the rules of mathematics! It is now obvious: you cannot play with the rules of mathematics, for they are not arbitrary; there are good reasons why they are known as *laws*!"

But wait a minute. Yes, the result is incorrect. But notice, we obtained this result by assuming that an equation from traditional algebra would also be valid in our imaginary algebra. But it need not be. The impression is essentially mistaken because we can construct a similar equation that represents the same relation. Specifically, we may establish a revised formulation:

$$|a^2| + |b^2| = |c^2|.$$

Here, the common operation of "absolute value" of numbers serves to convert them all to the same kind:

$$|9| + |{-16}| \overset{?}{\Leftrightarrow} |25|$$

$$9 + 16 = 25.$$

So our makeshift algebra now has an exact symbolic expression for the Pythagorean relation.

But you might still be unsatisfied. You might think that by extracting absolute values we have imposed an arbitrary trick to reproduce the correct result. Yet the traditional approach can also be said to be rigged to give the correct answer by involving the rule that the squares of negative numbers are positive. It would likewise work in an algebra where the square of every number is negative. Moreover, the use of absolute values is actually the very procedure that can be used to make the Pythagorean rule valid in the geometric interpretation of imaginary numbers. For example, consider a triangle having sides identified by the numbers, 4, $3i$, and $4 + 3i$ as in figure 8. If we apply the Pythagorean theorem directly, we obtain

$$(4)^2 + (3i)^2 = (4 + 3i)^2$$

$$16 + 9(-1) = 16 + 12i + 12i + 9(-1)$$

$$7 = 7 + 24i.$$

This result is incorrect. Why? The problem is that we assumed that complex numbers represent the lengths of lines. They do not. Remember that what mathematicians effectively represented with complex numbers was not length, but *direction*. Nonetheless, if

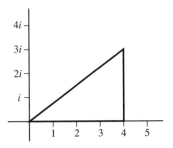

Figure 8. A triangle constructed on a plane of
positive and imaginary numbers.

we at least construe the positive and imaginary units along the reference axes as lengths, then we can try to express the length c of the hypotenuse by the equation

$$(4)^2 + (3i)^2 = c^2.$$

But here again, we find an incorrect result:

$$16 + 9(-1) = c^2$$

$$\sqrt{7} = c.$$

You can measure or estimate that the result should actually be 5. Accordingly, we can write

$$|(4)^2| + |(3i)^2| = |c^2|,$$

and, proceeding,

$$|16| + |-9| = |c^2|$$

$$16 + 9 = c^2$$

$$\sqrt{25} = c$$

$$5 = c,$$

we thus obtain the length of the hypotenuse. Thus even in traditional algebra it is sometimes necessary to introduce operations that help to attain a correct geometrical result.

   In any case, we can fashion artificial procedures to reproduce other valuable traditional results and relations. In particular, there is one outstanding property of imaginary numbers that we might like to somehow retain, in some form, in a mathematics that has no such things as imaginary numbers.

The multiplication of imaginary numbers involves a remarkable property that earned for the number i the occasional name of "rotation operator." For example, if we multiply 1 by i, and then multiply the result by i again, and so forth, we find that the results cycle through the same values:

$$1 \times i = i$$

$$i \times i = -1$$

$$-1 \times i = -i$$

$$-i \times i = 1.$$

So, after multiplying by i four times we return to 1. Likewise, any other number multiplied by i, successively, results in the same number. Since the results repeat after every fourth multiplication, mathematicians aptly represented such operations in terms of rectangular coordinate systems. For example, letting two straight lines cross one another perpendicularly, the resulting four lines could be subdivided as shown in figure 9. Here the values $1, i, -1, -i$, appear separated from one another by 90°, successively. Thus, a line segment originating at zero and extending to 1 might be "rotated" 90° by multiplying it by i. And, more generally, any vector line determined by a pair of real and imaginary coordinates might also be rotated in the same way. For example, the vector identified by the complex number $3 + i$ can be rotated counterclockwise 90° by multiplying it by i, and this operation might be carried out successively, as illustrated in figure 10. That is,

$$(3 + i) \times i = (3i + -1) = (-1 + 3i)$$

$$(-1 + 3i) \times i = (-i + -3) = (-3 - i)$$

$$(-3 - i) \times i = (-3i + 1) = (1 - 3i)$$

$$(1 - 3i) \times i = (i - 3(-1)) = (3 + i).$$

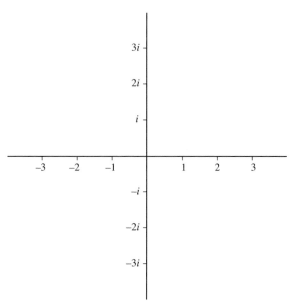

Figure 9. A representation of numbers as units on perpendicular lines.

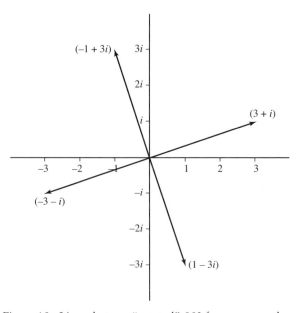

Figure 10. Lines that are "rotated" 90° from one another.

Mathematicians thus interpreted multiplication by $\sqrt{-1}$ as analogous to the operation of rotating a vector 90° in a rectangular coordinate system. This geometric interpretation helped to win acceptance for the validity of operations with imaginary numbers. It also revealed that imaginary numbers could be very useful in algebraic analyses involving cyclical functions, as in many aspects of physics.

So if we discard imaginary numbers, is there any way to establish an operation of rotation? Yes, and we have known how to do so especially since William Rowan Hamilton proposed his Theory of Couples.[20] One such way is to proceed as follows. Given any pair of numbers, (a, b), each being either positive or negative, we can define a variety of operations that affect them. For example, we can define an operation that changes their positions in the parenthetical expression

$$\leftrightarrow(a,b) = (b,a).$$

By charting the two numbers in a coordinate plane, we can give a geometrical meaning to this operation. For example, if the numbers are 1 and 3, we can write

$$\leftrightarrow(1,3) = (3,1),$$

and by interpreting these numbers as x and y coordinates, we may construct the lines shown in figure 11. By carrying out the same simple operation with several pairs of numbers, and drawing the corresponding lines, we can understand the general effects of the operation and hence give it a suitable name. In the same way, we can define a variety of other operations. For example, we can define an operation of "reflection":

$$-(a,b) = (-a,-b).$$

Likewise, we can find an operation that entails the same geometrical effects as multiplication by i. The operation consists of exchanging the positions of the two numbers, and then changing the sign of the first. Thus we define "rotation":

$$\circlearrowright (a,b) = (-b,a).$$

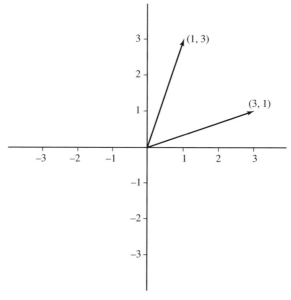

Figure 11. Lines determined by the same two numbers, but in different order.

By successively doing this same operation we obtain results that cycle through four values for the right term, and four for the left:

$$↻ (a,b) = (-b,a)$$
$$↻ (-b,a) = (-a,-b)$$
$$↻ (-a,-b) = (b,-a)$$
$$↻ (b,-a) = (a,b).$$

Again, we can interpret this operation geometrically by reading the terms in parentheses as coordinates. Given a vector determined by two numbers (x, y), we can rotate it 90° counterclockwise by exchanging the values of x and y and then changing the sign of x. To illustrate this, operate on the number pair $(3, 1)$:

$$↻ (3,1) = (-1,3)$$
$$↻ (-1,3) = (-3,-1)$$
$$↻ (-3,-1) = (1,-3)$$
$$↻ (1,-3) = (3,1),$$

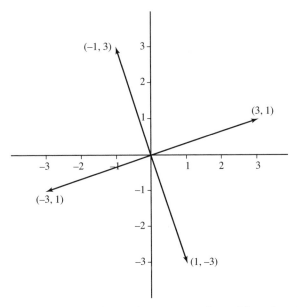

Figure 12. Lines "rotated" 90° from one another, without imaginary numbers.

and see how the results are charted (figure 12). The procedure is actually simpler than the procedure consisting of the multiplication of complex numbers. It is conceptually simpler too, and the symbolism corresponds better to the symmetry of the geometrical construction. Notice that it is immediately easy to define the reverse operation, that is, rotation in the clockwise direction. To rotate 90° to the right, exchange the values of the two coordinates and then change the sign of y. And finally, notice that there is no reason, here, to interpret any one particular number as "a sign of perpendicularity."

This demonstration should suffice to show how we can develop alternative methods for obtaining results that are usually obtained by traditional methods. It also shows how we can simplify mathematical procedures.

Now let us return to the question of how we ascertain algebraic equations that correspond to our fictitious arithmetical rules. As already demonstrated, we can test traditional algebraic equations to determine whether they agree with the stipulated rules, and if not, we can modify them accordingly. In this way we easily found a valid distributive relation for our fictitious multiplication rule and a general formulation of the Pythagorean theorem. Still, it is instructive to consider examples in which the process is not as direct. Given that early algebra developed from the study of finding solutions for quadratic equations, we may turn to such equations.

Traditional algebra tells us that the quadratic equation

$$(a + b)^2 = a^2 + 2ab + b^2$$

is valid for any mixture of positive and negative numbers. Let us see if it is valid in respect to our artificial arithmetic. As before, we test the equation by substituting numerical values. For example,

$$(5 + 3)^2 = (5)^2 + 2(5 \times 3) + (3)^2$$

$$(5 + 3)^2 = (25) + 2(15) + (9)$$

$$(8)^2 = 25 + 30 + 9$$

$$64 = 64.$$

So we see that the algebraic equation is valid in our artificial arithmetic, at least for combinations of positive numbers. But is it valid for negative numbers? For example,

$$(-5 - 3)^2 = (-5)^2 + 2(-5 \times -3) + (-3)^2$$

$$(-5 - 3)^2 = (-25) + 2(-15) + (-9)$$

$$(-8)^2 = -25 + 30 - 9$$

$$-64 = -4.$$

So we see that it gives an incorrect solution. The skeptic might now smile. However, knowing now at least one way in which it is not valid, let's see what went wrong. Notice that if only the term 30 had been negative then the result would have been correct. And why was it not negative? Because we multiplied $2(-15)$ in accord with our putative rules. But instead of placing the blame on the contrived multiplication rules, we can blame the algebraic equation. Indeed, we can easily solve the problem by making a slight change in the equation, so that now

$$(a + b)^2 = a^2 + ab2 + b^2.$$

And, given the same previous numbers, we obtain

$$(-5 - 3)^2 = (-5)^2 + (-5 \times -3)2 + (-3)^2$$

$$(-5 - 3)^2 = (-25) + (-15)2 + (-9)$$

$$(-8)^2 = -25 - 30 - 9$$

$$-64 = -64.$$

Perfect. And the new algebraic equation will work for *all* possible combinations of negative numbers. "Hum." If you actually felt somewhat pleased when you saw the previous incorrect result, you might now be a bit surprised.

So the problem was that by adopting the equation exactly as it is used in traditional algebra, we neglected to recognize that it might not be consonant with our artificial rules. In particular, the problem was rooted in the old assumption of the commutative "law" of multiplication. But, by applying instead the make-believe rules, that $+ \times - = +$ and $- \times + = -$, we found an easy solution. By the way, note that the initial difficulty we encountered stemmed purely from the combination of signs, not from the operation of squaring negative numbers and obtaining negative results.

According to our artificial rules, the equation

$$(a + b)^2 = a^2 + ab2 + b^2$$

is valid for all cases in which a and b are both negative. It is also valid for all cases in which a and b are both positive. It is also valid for all cases in which either a or b is equal to zero. But what about other cases? Consider a case in which a and b have different signs:

$$(5 - 2)^2 = (5)^2 + (5 \times -2)2 + (-2)^2$$

$$(5 - 2)^2 = (25) + (10)2 + (-4)$$

$$(3)^2 = 25 + 20 - 4$$

$$9 = 41.$$

This result is incorrect. To obtain the correct result, we might have expected to see

$$(3)^2 = 25 - 20 + 4$$

as one of the steps in the solution. But instead, not merely does the center term have the "wrong" sign, $+20$, now, but one of the squares, -4, also has the "wrong" sign. Moreover, notice that in the usual procedure the squares, 25 and 4, both are positive, so we may imagine that this demonstrates why, traditionally, "minus times minus is plus." Maybe now the skeptic smiles again.

Is there any way to modify the equation to obtain the correct result? Well, if, on the right side of the equation, we change the additions to subtractions, so that

$$(a + b)^2 = a^2 - ab2 - b^2,$$

we would hence obtain the correct result. But this approach might seem unsatisfactory. The earlier adjustment that led to correct results originated directly from taking into account the stipulated rules of multiplication; while, by contrast, changing the signs of the equation just to satisfy a particular case seems arbitrary. Better than just modifying algebraic equations as the need arises, we would also like to understand why the modifications are necessary. In the present example, we have altered the signs of the algebraic equation so that numerical values would correspond to the traditional procedure:

$$(^+5 + {}^-2)^2 = (25) - (10)2 - (-4).$$

Thus, the sum of two numbers of different signs, squared, is set equal to the sum of two positive squares and twice a negative middle term. Since then the two squares, 25 and 4, are both positive,

$$(3)^2 = 25 - 20 + 4,$$

even though one stems from a negative number, haven't we indirectly relied on the old rule that "minus times minus is plus"?

Instead of imitating the results of traditional algebra, let's elucidate the nature of the problem. To do so, we will use arithmetic and geometry to explain the origin of the usual algebraic expres-

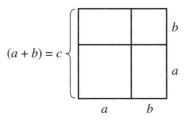

$(a + b) = c$

b

a

a b

Figure 13. Large square composed of two squares and two rectangles.

sions. We will then understand why early algebraists were led to conclude that "minus times minus is plus," that is, that all squares are positive. But we shall see that this old conception is not strictly necessary. Because we will demonstrate how to formulate problems about the relationships among squares in terms of our artificial rules.

Consider again the traditional equation

$$(a + b)^2 = a^2 + 2ab + b^2.$$

What does it mean geometrically? Well, if we assume that $a + b$ is greater than a and greater than b, we can illustrate the equation with a figure such as figure 13. This figure shows how a square having sides of length $(a + b)$ can be analyzed as the exact sum of two smaller squares and two rectangles. Or, say, it shows how a square having sides of length a can be expanded by adding b to its sides. We can give the figure arithmetical meaning by assigning numerical values to the lengths a, b, and c, as in figure 14. These

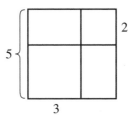

5

2

3

Figure 14. Assigning numbers to the lengths of the squares and rectangles.

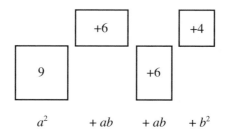

Figure 15. Adding the separate areas of each square and rectangle.

figures have areas determined by the product of their sides, for example, the area of a small rectangle here is $2 \times 3 = 6$. The largest square, $c^2 = 5 \times 5 = 25$, can be understood as the successive addition of the quadrilaterals in figure 15. Hence the square area "25" is constructed just by putting together the four quadrilaterals. There seems to be no ambiguity here.

Consider now a case in which areas are subtracted from one another. Given the same large square, with sides of length 5, suppose we reduce the length of each side, to find what will be the size of the remaining area. For example, what is $(5 - 2)^2$? Ordinarily, we might write

$$(5 - 2)^2 = (5)^2 - 2(5 \times 2) + (2)^2,$$

where the square is determined by the sum of two positive squares minus two rectangles. Thus, to summarize all such cases, traditional algebra employs the equation

$$(c - b)^2 = c^2 - 2cb + b^2,$$

or the equation

$$(c + b)^2 = c^2 + 2cb + b^2,$$

where c and b may have opposite signs. The operations in question may be represented geometrically (figure 16).

The relation among these areas can be expressed arithmetically:

$$(5 - 2)^2 = 25 - 10 - 10 + 4.$$

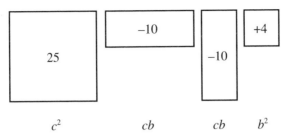

$$c^2 \qquad cb \qquad cb \quad b^2$$

Figure 16. The sum of two different squares minus two equal rectangles.

Notice especially that these formulations involve two *positive* squares, 25 and 4, to arrive at the square of (5 – 2). Hence, by interpreting (5 – 2) as ($^+5 + {}^-2$), as usual, we see how the notion that ($^+5)^2 = {}^+25$ suggests that, likewise, ($^-2)^2 = ({}^+4$).

Thus we can understand why early algebraists concluded that "minus times minus is plus." The reduction of a square, as demonstrated in a geometrical diagram, can be expressed algebraically as

$$(c - b)^2 = (c - b)(c - b),$$

and since the result

$$(c - b)^2 = c^2 - 2cb + b^2$$

involves the square of b as a positive quantity, it would seem that the square of any negative number is positive. Indeed, all squares would then seem to be positive. Accordingly, many algebraists, for over a thousand years, claimed that "there are no negative squares."[21] Some historians have thus surmised that even the Greeks, in their geometry, had assumed implicitly the rule that minus times minus is a positive.[22]

But that old inference is unjustified. We need not assume that the rule "minus times minus is plus" is the only way to translate the geometrical reduction of a square into arithmetic and algebra.

There is an asymmetry between the analysis of a square reduced by the subtraction of length from its sides and the analysis of a square enlarged by addition to its sides. Consider again the last diagram, wherein a square of area 25 is compared to two rectangles

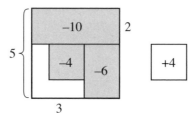

Figure 17. Taking too much away from the big square, and then adding some back.

of areas −10 each, and a smaller square of area 4. Observe what happens, in this case of subtraction, when we put together the areas geometrically. Representing the subtracted areas by shaded areas, we have figure 17. It illustrates that the reason why we need to add the square 4 is because by subtracting the two areas 10 and 10 we have subtracted too much. Thus, in such analyses of the reduction of squares we subtract *in excess* of what is necessary, only to then restore the excess. By contrast, in the analyses of the expansion of squares we add *exactly only* what is necessary to compose the large square. This asymmetric treatment leads to the *impression* that the square of a negative quantity must be positive. But we can well proceed otherwise.

We can, instead, subtract just the right amount. Consider the way of analyzing the square of $(5 - 2)$ shown in figure 18. This figure can be expressed arithmetically by

$$(5 - 2)^2 = 25 - 6 - 6 - 4.$$

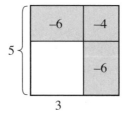

Figure 18. Taking away just as much as is needed to leave the medium-sized square.

Here we have a *negative* square: −4. The squares on the right side of the equation, +25 and −4, have opposite signs, just as the numbers on the left side, +5 and −2, have opposite signs. This symmetry is perfectly consonant with the rules + × + = + and − × − = −. Therefore, let us apply the artificial rules to describe this analysis.

Consider the following expression, where all the signs have been written explicitly for the sake of clarity:

$$(^+5 + {}^-2)^2 = (^+5)^2 + (^-2 \times {}^+3) + (^-2 \times {}^+3) + (^-2)^2.$$

Thus we obtain

$$(^+3)^2 = (^+25) + (^-6) + (^-6) + (^-4)$$

$$^+9 = {}^+9,$$

and this is the correct result. We can also obtain this result by means of this formulation:

$$(^+5 + {}^-2)^2 = (^+5)^2 + (^-2 \times {}^+3)^+2 + (^-2)^2,$$

which may be summarized algebraically as

$$(c + b)^2 = c^2 + ba2 + b^2,$$

where *a* and *c* are implicitly understood to be positive and *b* negative. Finally, note that since $a = (c + b)$ we might prefer to write

$$(c + b)^2 = c^2 + b(c + b)2 + b^2.$$

Notice that although we conceive an operation of subtraction of *b* from *c*, we use the expression $(c + b)$. Remember, according to our definitions, addition means only that numbers are to be combined. In the present case the "difference" results simply because the values to be substituted into *c* and *b* have different signs. Likewise, traditionally algebra also employs the additive equation

$$(a + b)^2 = a^2 + 2ab + b^2,$$

in cases wherein *a* and *b* have different signs.

Looking back at the diagrams we've used, you will notice that given one quadrilateral we have employed three others to construct a fifth quadrilateral. In the case of addition, given one

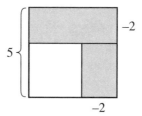

Figure 19. Taking away only two rectangles to leave the medium-sized square.

square, we add two rectangles and another square to construct a larger square. In the case of subtraction, we begin with a square and take away two rectangles and another square in order to arrive at a smaller square. But we can well simplify the construction and reduction of squares by employing four quadrilaterals instead of five. Figure 19 shows one way of reducing one square to a smaller one by subtracting only two rectangles. This diagram can be described by the equations

$$(^+5 + {}^-2)^2 = (5)^2 - (5 \times -2) - (5 - 2)(-2)$$

$$(5 - 2)^2 = (25) - (10) - (3)(-2)$$

$$(3)^2 = 25 - 10 - (6)$$

$$9 = 9.$$

Again, this result is correct. It may seem that here we have avoided the question of negative squares. But no, notice that we may also apply the same arithmetical description to cases in which the larger square is negative:

$$(^-5 + {}^+2)^2 = (-5)^2 - (-5 \times 2) - (-5 + 2)(2)$$

$$(-5 + 2)^2 = (-25) - (-10) - (-3)(2)$$

$$(-3)^2 = -25 + 10 - (-6)$$

$$-9 = -15 + 6$$

$$-9 = -9.$$

So this result is also correct. We may summarize such equations with the algebraic expression

$$(c + b)^2 = c^2 - cb - (c + b)b,$$

where again the expression $(c + b)$ indicates a difference simply because the values of c and b have opposite signs. This equation is yet another distinct way of representing the reduction of a square algebraically. Thus we see that we can ascertain various algebraic expressions corresponding to diverse geometrical analyses.

Now, looking back at all the arithmetical examples analyzed so far, you may notice that in every parenthetical expression, such as $(5 - 2)$ and $(-5 + 2)$, the greater quantity precedes the smaller. But what if we switch them, so that the smaller appears first? Are the same algebraic statements valid then? Considering our last formula, let $c = 2$ and $b = -5$, so that

$$(2 - 5)^2 = (2)^2 - (2 \times -5) - (2 - 5)(-5)$$

$$(-3)^2 = 4 - 10 + 3(-5)$$

$$-9 = 9.$$

This result is incorrect. Thus we can show that the algebraic expressions

$$(c + b)^2 = c^2 - cb - (c + b)b$$

$$(c + b)^2 = c^2 + b(c + b)2 + b^2$$

$$(c + b)^2 = c^2 + ba2 + b^2$$

are not valid when b is greater than c. But why, you may ask, are these expressions only valid when c is the greater? The answer is found in the original conditions involved in our first geometrical diagram. Remember that we specifically assumed that c is greater than a and greater than b. Accordingly, we obtain impossible results when we violate those features of that diagram.

The examples provided should suffice to illustrate how we can formulate an algebra corresponding to whatever arithmetical rules we establish. Again and again it first seemed that by changing basic rules we would be left with a crippled algebra lacking important tools: the distributive rule, the Pythagorean theorem, the rotation operator, quadratic equations. Yet we were able to develop a working system that included such tools.

New mathematical systems can have both similarities and differences relative to older mathematics. The sample system currently developed admits as valid certain expressions that are also valid in traditional mathematics, while it excludes some others. It also admits certain expressions that are not valid in traditional mathematics. We have seen that this artificial algebra requires that certain various arithmetical and geometrical relations be expressed by *distinct* algebraic expressions. Meanwhile, in traditional algebra many individual algebraic expressions apply to broader varieties of relations indiscriminately. Therefore, compared to traditional algebra, the artificial algebra will be more complicated in general, especially in problems involving multiplication of numbers with different signs.

Furthermore, you may notice that some of the rules of this artificial algebra, which deviate from standard algebra, resemble rules in some other branches of mathematics. For example, consider the artificial rule that the product of two oppositely signed numbers bears the sign of the first number. It seems to share with vector algebra the following algebraic form:

$$(a \times b) = -(b \times a),$$

if at least we let a and b designate vectors, magnitudes having different directions. Hence, maybe it might seem pointless to use an allegedly new rule if it just serves to replicate an algebraic form that is already available within familiar methods. But upon close inspection, we may find that, although some cases of some rules seem to be equivalent in two algebras, the rules and products also involve subtle differences. Actually, the rule above does not really have the same meaning in vector algebra as in the artificial algebra. Because, in the latter, the product is not perpendicular to a plane determined by a and b. In the artificial algebra the product is just a signed number, which might be interpreted as having the "direction" of either a or b, but not a third kind of direction.

Moreover, even if a new algebra includes rules that are exactly equivalent to some preestablished mathematics, that does not render the newer system useless. For instance, if we argue that the artificial algebra in question is useless to the extent that it seems to replicate vector algebra, then we should remember that vector algebra originally gained acceptance precisely *because* it served to replicate results that could be obtained with coordinates and analysis. It would be inconsistent to argue that newer mathematics are likely useless insofar as they mirror older mathematics, while, at the same time, valuing the usual methods as useful precisely because they correspond to one another.

To properly appraise a novel algebraic system, hence, one has to study its properties in detail. To be sure, the features that might first strike the eye are those that deviate, in a seemingly unnatural way, from the forms of familiar mathematics. Thus it is easy to look at the artificial scheme outlined in this section and identify its oddities. But some oddities are less unpleasant than others.

This makeshift system also introduces simplifications. For example, the arithmetic underlying the algebra will be no more complicated than traditional arithmetic, and it is in many respects simpler. Moreover, many operations involving radicals and signless exponents will be simpler. To highlight this advantage, consider the following properties of the contrived mathematics. First, the extraction of any root of any number yields a single solution:

$$\sqrt[x]{a} = unique.$$

In traditional algebra there are as many roots to each a as the value of any given x. The univocal solution in the improvised rule eliminates all asymmetries, complications, and ambiguities stemming from the traditional multiplicity. Second, in the artificial scheme the operation of raising a number to a power and the operation of extracting its corresponding root are opposite operations; each exactly undoes the effect of the other:

$$\sqrt[x]{(a^x)} = a \quad \text{and} \quad (\sqrt[x]{a})^x = a,$$

(even when a and x have the same sign). Strictly speaking, in traditional mathematics this is not true; for example, $4^2 = 16$, but

$\sqrt[2]{16} = \pm 4$. Third, the operation of raising a number to a power and the operation of extracting a root are commutative:

$$\sqrt[x]{(a^y)} = (\sqrt[x]{a})^y.$$

This relation is not strictly valid in traditional mathematics owing to the multiplicity of solutions for the extraction of roots. Fourth, the operation of raising a number to a power and the operation of changing its sign are commutative operations:

$$-(a^x) = (-a)^x,$$

while traditionally they are not. Fifth, the extraction of roots and the operation of changing the sign are commutative:

$$-\sqrt{a} = \sqrt{-a}.$$

Traditional mathematics does not have this property. Sixth, the extraction of square roots is distributive with regard to multiplication:

$$\sqrt{(ab)} = \sqrt{a} \times \sqrt{b}.$$

Here this property is true, without any ambiguity. By contrast, in traditional mathematics, as usually understood, this property is said to not be true for negative numbers.

In addition to such advantages, the artificial arithmetic and algebra of logarithms also introduces simplifications. There too the operation of obtaining a logarithm will involve a unique solution for any number z to any base x:

$$\log_x z = unique.$$

Moreover, the artificial math involves an elegant correspondence among the three operations: raising a number to a power, extracting the corresponding root, and obtaining the logarithm:

$$x^y = z, \quad \sqrt[y]{z} = x, \quad \log_x z = y.$$

These equations are all simultaneously true for all cases in which x and z have the same sign. Again, traditional algebra lacks this property.

There are many aspects of the sketchy mathematics that still remain entirely undeveloped. For example, how should we define the operation of division? Or, say, how should we deal with fractional exponents? Furthermore, how can we harmonize the artificial algebra with preestablished branches of mathematics, such as trigonometry? Or how does it harmonize with coordinate geometry (this particular question will be discussed in chapter 7)? Yet such questions need not be answered here, since the main goal of the inquiry was simply to demonstrate how we can design mathematical principles and elucidate their consequences. In doing so, we did not encounter any contradictions that were not merely apparent. Thus we have demonstrated some of the arbitrariness in traditional mathematical principles. So, to a question such as "Can minus times minus be minus?" the answer is yes, an old though often denied and neglected lesson from the history of mathematics; "the choice of sign-rule is the user's—a fact not fully recognised by mathematicians generally until the 19th century."[23]

In conclusion, we have seen how we can modify the "laws" of mathematics to effectively devise new mathematics based on rules that we may desire for one reason or another. Unfortunately, every new rule or property we introduce might entail difficulties that may need to be solved or elucidated. Whether one mathematical system is simpler than another, overall, is a difficult question to answer, especially if the newer system is incompletely developed, and if the earlier system is not compared in its entirety to the new. Whether an unnatural mathematics has any practical value can be ascertained by attempting to apply it to particular problems. Whichever mathematics one chooses to employ or develop is a matter that depends on convenience, generality, and effectiveness in solving specific problems. It is also a matter of habit.

For our purposes, merely experimental, we have devised an artificial system that is by no means a full-fledged mathematics, nor even an alternative to ordinary math; it remains merely a suggestive skeleton, a mannequin. But at least we see that we have considerable freedom to devise new numerical algebras that, someday and for some purposes, might be preferable to those presently available.

UNITY IN MATHEMATICS

Before proceeding to discuss how to formulate mathematical rules that correspond to practical notions, we should discuss whether distinct mathematical systems can coexist. For example, in one algebra minus times minus makes plus, and in another, minus times minus makes minus; must one or the other be wrong?

Two statements seem mutually contradictory: $-2 \times -2 = +4$ and $-2 \times -2 = -4$. We establish, by definition, that $+4$ and -4 are not equal things, that instead, they are opposite. Hence we cannot have $+4 = -4$, and they are not different names for the same thing. On a first acquaintance it might seem that the two statements are necessarily contradictory. That they cannot both be true. It might seem that we have to discern which one, if either, is correct. But remember that we are dealing with symbols here. To some extent, if we are free to assign or modify the meaning of such symbols we may well find ways to harmonize the two statements.

Notice, for example, that we are used to not sensing any conflict between the propositions $\sqrt[2]{4} = 2$ and $\sqrt[2]{4} = -2$, which, if we didn't know better, might seem to suggest the contradiction $+2 = -2$. So perhaps one way to harmonize $-2 \times -2 = +4$ and $-2 \times -2 = -4$ would be to say that the operation of multiplying negative numbers produces two results. But maybe you don't like this. You may say that the traditional concept of multiplication is just too well established and widespread to modify or extend its meaning. So maybe you much prefer the usual notion that multiplication yields unique solutions. You accept double solutions in the square root operation, but you deny them for multiplication.

Okay, we need not change the traditional concept of multiplication. If so, is there any other way to harmonize the two statements in question? Yes. One way is to establish that the $\times$ sign in "$-2 \times -2 = -4$" is not multiplication really, but that it should actually mean something different, though similar. For example, let us call it "replication," and change the sign so that we write

$$-2 \times -2 = 4 \quad \text{and} \quad -2 \otimes -2 = -4.$$

Now we simply have two distinct operations telling us how to interrelate negative numbers according to certain rules. Still, this raises new issues. Whether or not -2×2 is equal to $-2 \otimes 2$ is a question that we must now decide depending on what meaning we choose to assign to the $\otimes$ sign. If we follow the rules suggested in the previous section, then $-2 \times 2 = -2 \otimes 2$, but $2 \times -2 \neq 2 \otimes -2$. Unlike multiplication, replication would not be a commutative operation. As suggested previously, we have quite some latitude in formulating a new symbolic algebra so long as we keep its consequences mutually consistent.

Still, we may yet wonder: What if early algebraists such as Cardano and Harriot had convinced most other mathematicians that minus times minus makes *minus*, would we now call that "multiplication," and what would we call the deviant rule minus into minus makes plus?

Maybe we can try to supersede the historical contingency of words by conceiving of different "kinds" of multiplication: say, multiplication1, multiplication2, etc., where the numbering corresponds to the order in which they were conceived (or widely recognized as plausible), and we can even give them new proper names such as "multiplation," "multimation," (silly as these names may sound), etc. If so, which one is the "real" multiplication? Perhaps the answer is arbitrary: we decide what we mean by the word multiplication.

Let's consider a few relevant examples from history. In some ancient civilizations, zero and one were not considered numbers; hence in the corresponding rudimentary algebra, the statement

$$a \times b > a$$

was true for "all" numbers. But once one and zero were admitted as numbers, multiplication was reconceived as satisfying the statement

$$a \times b \geq 0$$

for all numbers. Back then, before negative quantities were accepted, it was ridiculous and meaningless to multiply quantities smaller than nothing. But once they were accepted, the meaning of "multiplication" came to be modified by extension. It no longer necessarily implied an increase in quantity, as imaginaries, negatives, and even zero and one, were admitted as numbers. One could then say that in some cases $a \times b \geq 0$, but not in others. Still, all multiplication had the commutative property

$$a \times b = b \times a,$$

as an intrinsic part of its meaning. But once Hamilton conceived of a new kind of number, this was no longer a general necessary property of multiplication. This radical step seemed repulsive to some people at the time. Still, mathematicians could hence say that multiplication is always defined partly by an associative property:

$$(a \times b) \times c = a \times (b \times c).$$

This property was valid for quaternions, vectors, imaginaries, negatives, positives, and more. But once Gibbs and Heaviside modified the rules of vectors, to create a new vector theory independent of quaternions, multiplication was then not associative with respect to such vectors. One could then say that in some cases $(a \times b) \times c = a \times (b \times c)$, but not in others.

What happened then? Did people finally believe that they had so much extended and modified the concept of multiplication as to deem it self-contradictory? No, the path taken then was to interpret vector "multiplication" as a distinct operation, and call it the "cross product." Likewise, vectors, which had been construed as the imaginary part of any quaternion, came to be construed as a geometrical magnitude rather than a kind of numerical element. Hence we see how we can modify and distinguish concepts in such a way that seemingly contradictory notions can be simultaneously admitted without actually generating contradiction.

But note that we sometimes impose certain restrictions to make various concepts cohere. For example, since $\mathbf{a} \times \mathbf{b}$ is not really the "multiplication" of vectors but their cross product, then what happens if we nonetheless *multiply* the vectors? The restriction here is to say that we cannot really multiply vectors, that we either obtain their cross products, or multiply their lengths irrespective of their directions, and so on. A similar difficulty happens with other operations when they are applied to new mathematical "objects" (such as vectors). For instance, ask your math or physics teacher what is the result of dividing two vectors. Many teachers will answer that there is no result; that you cannot divide vectors. On this matter, the physicist and mathematician Banesh Hoffmann cautioned:

> we naturally ask ourselves how we can *multiply* and *divide* vectors. But actually there is no reason—except perhaps misguided optimism—why we should expect to be able to do so usefully. For example, though we can "add" colors (as on a color television screen) or "subtract" them (as in amateur color photography), it does not occur to us to ask how we could multiply or divide colors by colors. Again, we can add and subtract money; and if we are astute and lucky, we can multiply money—but by numbers, not by money. If we start with $100 and triple our money, we end up with $300 which is $100 multiplied by 3, not multiplied by $3 . . .[24]

So traditional operations need not apply to all mathematical "objects" in like way as they do not necessarily apply in all practical contexts. Yet we can devise new operations that resemble traditional ones, and thus establish some limited meaning for expressions such as "vector multiplication," and this can all harmonize with preestablished mathematics.

Still, even if we admit a distinct new operation, we can find, on close inspection, that it is more different from the traditional operation from which it stems than we may have assumed. For example, while traditional multiplication of real numbers always results in real numbers, the cross product of vectors does not produce another vector (unidirectional line segment), strictly speaking, but something less definite, that is, a "pseudovector." Thus new operations and mathematical objects might engender other new concepts. Accordingly, Hoffmann commented:

> Perhaps it worries us a little that there are different kinds of vectors. Yet we have all, in our time, survived similar complications.

Take numbers, for example. There are whole numbers and there are fractions. Perhaps you feel that there is not much difference between the two. Yet if we listed the properties of whole numbers and the properties of fractions we would find considerable differences. For instance, if we divide fractions by fractions the results are always fractions, but this statement does not remain true if we replace the word "fractions" by "whole numbers." Worse, every whole number has a next higher one, but no fraction has a next higher fraction, for between any two fractions we can always slip infinitely many others. Even so, when trying to define *number* we might be inclined to insist that, given any two different numbers, one of them will always be the smaller and the other the larger. Yet when we become more sophisticated and expand our horizons to include complex numbers like $2 + 3\sqrt{-1}$ we have to give up even this property of being greater or smaller, which at first seemed an absolutely essential part of the idea of number.[25]

Happily, such complications, if pursued systematically, can lead to the development of new mathematics that can be quite valuable and even practically useful. Hence, Hoffmann suggested: "The best thing to do is to keep an open mind and learn to live with a flexible situation, and even to relish it as something akin to the true habitat of the best research."[26]

Returning to minus times minus, let's review some of the ways in which seemingly contradictory rules can be reconciled. First, one might decree that a preestablished operation has a new extended meaning, for example, that multiplication yields double solutions. Otherwise, we might distinguish the deviant rule "minus times minus makes minus" as a different operation that is not multiplication at all. Or we might say that it is a different kind of multiplication. Or a complementary approach would be to say instead that the operation is always multiplication but that the "negative numbers" in $-2 \times -2 = -4$ are not really negatives, but a different kind of number that should be assigned a new sign. Or we could even say that the operation is *not* multiplication and that neither are the numbers "negatives." Then we might still wonder: what happens to such new numbers if we multiply them? In any case, the point is that as we devise new mathematical rules we can often reconcile them with preexisting rules so long as we are willing to modify certain concepts.

Moreover, we can simply say that we have devised distinct independent algebras irrespective of whether they mutually cohere or not. Various symbolic algebras do not need to all be different ways of stating the same things. But even in this approach, once we devise various algebras we can interrelate them by ascertaining a more general algebra that describes them. For example, if we have one algebra where minus times minus makes plus, and another where it makes minus, they both may yet be described by a more general algebra characterized by what is common to both: that multiplication yields only one solution, whatever its sign. So there are ways to systematically unify diverse algebras.

An important consequence of the now old realization that we can devise new mathematics is that we now know that any one direction pursued in mathematics is not necessarily pursued at the expense of another. And even earlier, many mathematicians who characterized negative and imaginary expressions as "impossible" nonetheless admitted such expressions for various reasons. A major reason was to attain greater generality in algebra. Another was to employ algebra to study the limits of what is possible, as argued, for instance, by John Playfair: "No part of the language of algebra, it is plain, can be regarded as of greater importance than that in which these imaginary characters are employed. It explains the nature of those limits by which the possible relations of things are circumscribed, and marks out the conditions that are capable of being united in the same thing, or in the same system of things."[27] Accordingly, a reason why we have focused on abstract expressions as a vehicle for introducing the concerns of meaning in mathematics is because traditionally it was with such expressions that the distinctions between what seemed more or less possible were made.

Meanwhile those who wanted algebra to be more "scientific" or who admitted only meaningful or possible relations, such as

Francis Maseres, Robert Simson, William Frend, Lazare Carnot, and others, overtly rejected imaginary expressions. Perhaps some thought that algebra should only be developed in one direction, and thus sought to restrict its elements and methods. Playfair replied:

> Among certain *Purists* in algebraic language, no quarter is allowed to such modes of expression as we have been here treating of; and the investigations that proceed by them are considered as delusive artifices, unworthy of the name of science. To this opinion, however, we can by no means subscribe. Whatever has served for the discovery of truth, has a character too sacred to be rashly thrown aside, or to be sacrificed to the fastidious taste of those who make truth welcome only when it wears a particular dress, and appears arrayed in the *costume* of antiquity.[28]

Again, even if imaginary expressions lacked evident meaning, they were yet highly useful for obtaining a wealth of interesting results. Many results were admitted as "truths" because they were confirmed by other methods, so even in physical applications many abstract expressions became useful instruments of calculation and analysis. Thus Playfair characterized impossible expressions as "valuable instruments in the discovery of truth. The anticipations they afford are of infinite value; and no man who knows the importance even of scientific conjecture, will willingly give up the advantage to be derived from them."

Likewise, when Hamilton formulated his Theory of Couples, to show that he could reproduce mathematical relations commonly obtained with imaginary numbers *without* using such numbers at all, mathematicians did not hence abandon or renounce imaginary and complex expressions. Instead, the algebra of couples became just an alternative method for investigation and analysis. And likewise, when Gibbs and Heaviside changed the rules of vectors to formulate a new theory, the preestablished vector rules of the quaternion theory continued to be employed by mathematicians and physicists who preferred them. There are many instances when leading mathematicians have tried to fundamentally change or restrict mathematics, but instead have succeeded only in the creation of new mathematics, not in replacing the old.

Enough has been said to illustrate the plasticity of symbolic concepts and show that diversity can exist in mathematics. We have learned that creativity in mathematics is not necessarily cultivated at the expense of preestablished concepts and procedures.

Making a Meaningful Math

Under the banner of symbolical algebra, we can devise new systems of rules on how to manipulate arbitrary symbols. Like making the rules of a game. It doesn't matter if the symbols have no meaning. Such symbols need not even represent numbers. We can define operations that establish the rules of combination among symbols, and perhaps equivalence among certain symbols or groups of symbols. We can devise operations that serve to transform some symbols into other symbols. Admittedly, few people might be interested in such arbitrary systems of rules. Unless, maybe, we happen to use such systems to construct some impressive patterns or structures, or to get some useful results. It would be rewarding to find that a seemingly arbitrary system of rules and symbols happens to work for some practical purpose. But it would be strange to willfully pursue a meaningless game of symbols, exhaustively, in hopes that eventually useful meaning will emerge.

Instead, we can also design meaningful symbolical algebras. We can define symbols on the basis of what we want them to represent. Symbols can stand for objects, processes, thoughts, relations among them, and whatnot. We can define operations among such symbols to describe interrelations, operations, or transformations among the things to which the symbols refer.

Thus there are many ways that meaning can function in mathematics. In some cases, meaning can exist before symbolical rules and conventions are established, and thus such rules and conventions can be contrived to obey that meaning. In other cases, the symbolic structures and rules happen to be already available, and we then proceed to try to associate a particular or general meaning with them.

Now, in the sections ahead, we will consider examples of how to devise meaningful algebras. First we will consider ways in which we can ascertain some plausible meaning in preestablished algebraic forms. Afterward we will consider examples of how to devise and develop numerical concepts that correspond to a predetermined meaning.

Many mathematical practices originate from daily physical experience. Ordinary activities of counting, measuring, and describing forms and regularities give rise to mathematical concepts. And although the growth of mathematics overall is not restricted by practical needs, and it need not be, we can nevertheless formulate special mathematics that accord with practical experience.

FINDING MEANING

In chapter 6 we devised a symbolic algebra, guided mainly by formal concerns. We sought symmetry and arithmetical simplicity. Of course, the exercise was motivated indirectly by ordinary empirical considerations, such as by basic concepts of motion, areas, and the notion of symmetry between left and right. So now we will illustrate some ways in which those artificial symbolic rules can be applied in more meaningful contexts.

We established, in particular, that $-a \times -a = -a^2$ and $\sqrt{-a^2} = -a$. If these rules seem weird and deviant, you might maybe imagine that if they can be applied to any physical contexts at all, then it will be in the attempts to account for strange and counterintuitive things and relations in the invisible structure of the world. But not necessarily, no. At an introductory level, practical mathematics involves the labor of describing, by means of symbols and rules, physical operations and relations that are easily observable.

We already discussed a few examples in which our experimental rules of signs can be interpreted geometrically. The following additional examples illustrate other simple ways in which an algebra involving such rules can be applied to some of the usual contexts where signed numbers are ordinarily meaningful.

Consider the context of assets and debts. Let positive numbers represent assets and let negatives represent debts. The square root of $25.00 is $5.00. But what about the $\sqrt{-25.00}$? By the traditional rules, the operation results in the imaginary pair $\pm 5i$, but since this result is meaningless we routinely and tacitly reinterpret the problem as

$$-\sqrt{|-25.00|} = -\sqrt{25.00} = -1 \times 5.00 = -5.00,$$

and we say that the debt is $25.00, and its square root is $5 owed. Thus traditional algebra does not give the result immediately; we employ several operations. We know what the result should be, so we carry out the combination of rules that will give the correct result, or we solve the problem by formulating it with the help of words. Instead, by using the alternative algebra we can now simply write $\sqrt{-25.00} = -5.00$. Here a single operation gives the correct result directly. This operation is as simple as with the case of assets.

Analogous considerations apply also in the analysis of motion. In this context we use positive numbers to represent, say, motions to the east, and negatives to represent motions to the west. Suppose we want to use an algebra as a language for communicating instructions on how many steps to move and in what direction. To tell someone to move twenty-five steps to the east we may write +25. And what if we then say "now move the square root of that displacement"? If we write $\sqrt{+25}$ the result is ± 5, an impossible result because the person cannot simultaneously move in the two directions. One of the results is equivocal, namely, -5, so we must disregard it. Since we want the exact physical solution, we can ask for the principal square root: $\sqrt{+25} = +5$. This result conveys the number of steps and the right direction. Meanwhile, to convey

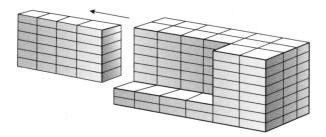

Figure 20. Removing a section of twenty-four blocks.

motion of twenty-five steps west we may write -25. "Now move the square root of that displacement." With the traditional rules $\sqrt{-25} = \pm 5i$, so again we reinterpret this instruction as "move west the principal square root of the *distance* traveled," because the extraction of the root of the *displacement* gives the wrong result. The distance is $|-25|$ so we write $-\sqrt{|-25|} = -5$. Again, if we instead use the alternative rules $\sqrt{-25} = -5$ and $\sqrt{+25} = +5$, we obtain the exact results directly.

Another context in which we can make sense of the rules in question is the analysis of areas and volumes. Consider a collection of blocks occupying a certain volume. Let positive numbers represent additional blocks adjoined to the collection, and negative numbers represent blocks removed. Suppose we remove a horizontal row of four blocks; we may express this operation by -4. Or suppose we remove a vertical column of six blocks; let that be -6. Accordingly, if we remove a rectangular section of blocks framed by four blocks on one side and six on the other (figure 20), we can express this operation by $-4 \times -6 = -24$. Likewise, we can analyze the removal of blocks along three dimensions, width, height, depth, that is, we can analyze the removal of blocks in terms of cubic volume. In this case both algebras give the same results directly because in both algebras $-a \times -a \times -a = -a^3$. Thus we may represent operations concerning areas and volumes by using various rules of signs.

We may also represent notions of position. As usual, we often use positives and negatives to designate positions of points along a number line, that is, to the right or left of zero. With the alternative rules we can also represent the positions of figures relative to

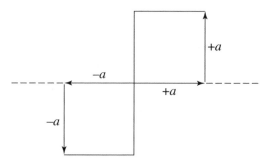

Figure 21. Opposite trajectories that determine opposite squares.

a reference line. Consider a straight dotted line as a reference axis, and from a point on it trace two lines of equal length in opposite directions along it: $-a$ and $+a$. And on the distant extremities of each line trace two perpendicular lines described again by $-a$ and $+a$. The two pairs of lines determine two squares, shown in figure 21. In traditional algebra $+a \times +a = +a^2 = -a \times -a$. But according to the alternative rules, we have instead $+a \times +a = +a^2$ and $-a \times -a = -a^2$. This description tells us that in a sense the two squares are opposite; specifically, that they are positioned on opposite sides of the reference axis.

Similar considerations can lead us to find meaningful applications of sign rules in the description of directed lines, that is, vectors. Again take two motions in opposite directions, $-a$ and $+a$, and then two perpendicular motions, $-b$ and $+b$. These lines can be taken to determine two triangles, where the hypotenuse of each is given by the vector sum of their sides (figure 22). Ordinary algebra tells us that

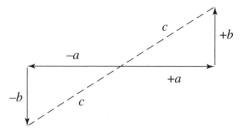

Figure 22. Opposite vectors that determine two triangles.

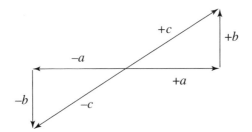

Figure 23. Opposite vectors that determine two triangles and two more opposite vectors.

$$\sqrt{(+a)^2 + (+b)^2} = c = \sqrt{(-a)^2 + (-b)^2},$$

whereas the alternative rules give

$$\sqrt{(+a)^2 + (+b)^2} = +c \quad \text{and} \quad \sqrt{(-a)^2 + (-b)^2} = -c,$$

so we obtain an exact and direct symbolic representation of the pair of vectors, which, like the other pairs, are equal in magnitude but *opposite* in direction (figure 23).

Another context wherein various algebras can usefully be applied is the representation of curves. The development of coordinate geometry since the days of Descartes was a major force in the growth of algebra and its application to the physical sciences. By correlating lines with equations, scientists represented and analyzed shapes of bodies, paths of light, trajectories of projectiles, motions of planets, wave motions, and more.

Since we have devised an algebra based on modifications of traditional rules, what are some of the corresponding effects on coordinate geometry? We may keep the usual procedures on how to plot points on a plane having two mutually perpendicular reference axes X and Y. We then find that some equations generate the same line in the two algebras. For example, the equation $y = x$ corresponds to the same line in the artificial algebra as in ordinary algebra (figure 24). Likewise, the equation $y = x + b$ has the same effect in both algebras.

However, many other equations correspond to different lines in the two algebras. For example, the equation $y = x^2$ when inter-

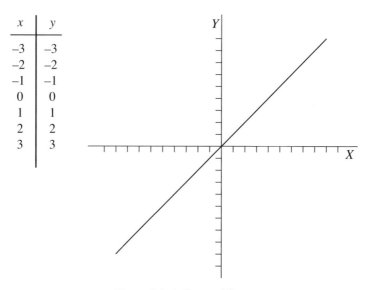

x	y
−3	−3
−2	−2
−1	−1
0	0
1	1
2	2
3	3

Figure 24. A diagonal line.

preted in the traditional algebra corresponds to the curve of a parabola (figure 25). Note that each branch of the parabola is an exact reflection of the other with respect to the Y-axis. Mathematicians call this kind of symmetry "bilateral." Yet the same equation, $y = x^2$, interpreted in the artificial algebra yields a different curve (figure 26). This curve does not have bilateral symmetry, but it does have another kind of symmetry. Notice that if the upper branch of this curve is reflected with respect to the Y-axis, and then the resultant is reflected with respect to the X-axis, we obtain the lower branch. Thus the figure can be rotated 180° in either direction and it continues to represent the same figure. This is called rotational symmetry. Hence we see how one equation can generate different curves when interpreted in coordinates according to the rules of one algebra or another.

Another example: in traditional algebra if we consider the equation $y = \sqrt{x}$ we obtain the curve shown in figure 27. Note that there are no points charted for negative values of x, because, since the square roots of negative numbers are imaginary, they have no

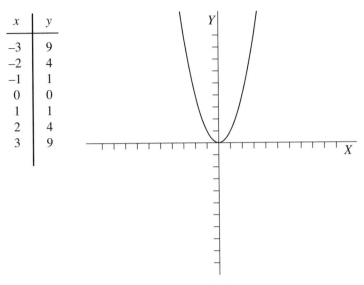

x	y
–3	9
–2	4
–1	1
0	0
1	1
2	4
3	9

Figure 25. A parabola.

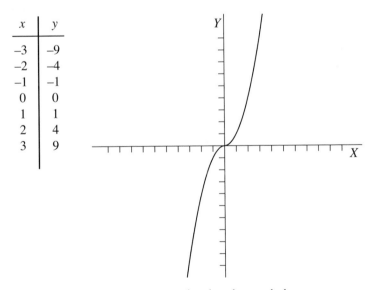

x	y
–3	–9
–2	–4
–1	–1
0	0
1	1
2	4
3	9

Figure 26. A curve related to the parabola.

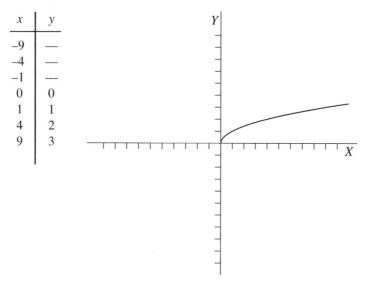

x	y
−9	—
−4	—
−1	—
0	0
1	1
4	2
9	3

Figure 27. A discontinuous curve.

place in the plane of positive and negative coordinates. By contrast, the same equation $y = \sqrt{x}$ in the manufactured algebra produces the curve shown in figure 28. The resulting curve is complete and symmetric.

Further example: in traditional algebra the equation

$$y = x^3 + 2x^2 - 8x$$

describes the curve of figure 29. Meanwhile, if we solve the same equation for values of x according to the contrived rules we get a different curve. For example, for the value $x = -3$, we solve the equation according to the rules previously described, in this manner:

$$y = (-3)^3 + 2(-3)^2 - 8(-3)$$

$$y = -27 + 2(-9) - (24)$$

$$y = -27 + 18 - 24$$

$$y = -33.$$

(Remember that in the make-believe rules the multiplication of

x	y
–9	–3
–4	–2
–1	–1
0	0
1	1
4	2
9	3

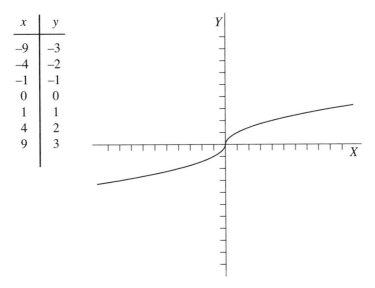

Figure 28. A continuous curve.

x	y
–4	0
–3	15
–2	16
–1	9
0	0
1	–5
2	0
3	21

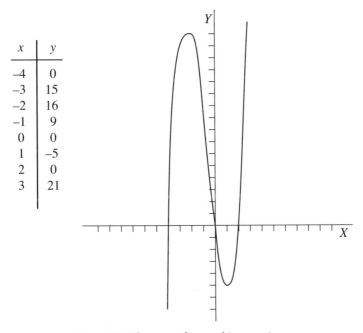

Figure 29. The curve for a cubic equation.

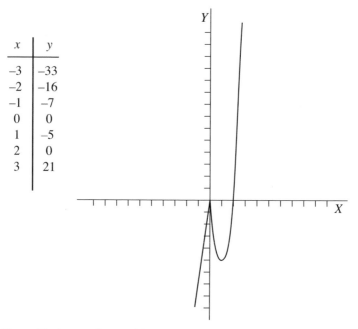

x	y
–3	–33
–2	–16
–1	–7
0	0
1	–5
2	0
3	21

Figure 30. A curve for a cubic equation in a noncommutative algebra.

signed numbers replicates the sign of the first multiplier.) Carrying out the same procedure for other values of x, and plotting each point, we obtain the correlated values and corresponding curve shown in figure 30. This curve may seem somewhat less graceful than the previous one. But such aesthetic considerations do not necessarily imply any shortcoming of the artificial algebra. An equation that describes a "nice" curve in one algebra might describe a different curve in another. In any case, one advantage of the artificial algebra under consideration is that we can use it to modify curves in ways that are not available in ordinary algebra. For example, in the artificial algebra, multiplication is not commutative, and this property can be applied usefully in the construction of curves. In the traditional algebra the following expressions all represent exactly the same algebraic relation:

$$y = x^3 + 2x^2 - 8x$$

$$y = x^3 + x^22 - x8$$

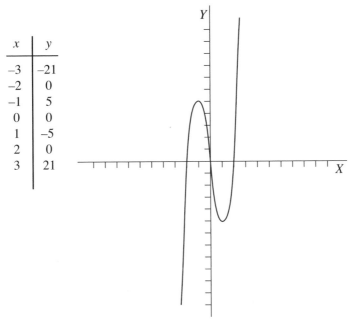

x	y
-3	-21
-2	0
-1	5
0	0
1	-5
2	0
3	21

Figure 31. A symmetric curve for a cubic equation in a noncommutative algebra.

$$y = 1x^3 + x^22 - x8$$

$$y = 1x^3 + 2x^2 - x8.$$

Thus, in traditional analytic geometry, these four equations describe only one and the same curve, which was illustrated before. In the artificial algebra, however, these equations represent distinct curves. For example, by solving the equation $y = x^3 + x^22 - x8$, for multiple values of x, and then plotting each point we obtain figure 31. This wave curve is symmetric. By contrast, the curve generated by the traditional rules, as in figure 29, is not. The potential value of a noncommutative algebra applied to coordinate geometry is formidable: we are then be able to construct a great variety of curves that otherwise have no algebraic representation. Most any equation that ordinarily represents but a *single* curve in traditional coordinate geometry can henceforth, by rearranging the order of its variables and coefficients, come to represent *sev-*

eral curves. Even if this were the only useful by-product of our artificial game of symbols we would yet have to judge our symbolic experiments as having been remarkably fruitful.

We can continue indefinitely the task of ascertaining meaningful applications of the symbolic rules of whatever algebra we choose to investigate. But enough has been said to illustrate, by way of introduction, the way in which plausible symbolic rules can be applied to the description of figures, positions, motions, monetary amounts, and so forth.

DESIGNING NUMBERS AND OPERATIONS

Now we will illustrate the task of developing new mathematical representations of relations among ordinary things. That is, rather than apply to experience rules that were established beforehand, we will develop new concepts and rules based on empirical notions. To do so, let's analyze usual elementary descriptions of some simple physical notions. By elucidating the divergence between traditional mathematical accounts and the relations they seek to describe, we can formulate mathematics that represents phenomena more exactly.

Consider the notion of length. We say that the length of an object is the distance between its extremities. Mathematically, a length may be described as the distance or separation between two stationary points along a reference line. Thus we apply reference systems, composed of straight lines divided into units, to describe numerically the length of an object. For example, we can align a rectangular reference system with a stationary rod such that the length of the rod is oriented parallel to a reference axis (figure 32). By reference to the X-axis, the two extremities of the rod are separated by a distance of four units of length. Arithmetically, this distance can be expressed in many ways, such as

$$4, \quad 2+2, \quad 3+1, \quad 4+0,$$
$$1 \times 4, \quad 7-3, \quad 9-5, \quad 8 \div 2.$$

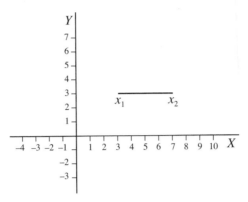

Figure 32. A straight rod aligned with an axis of
a coordinate system.

Algebraically, this same distance can be expressed in many ways
also. For example, let constant numerical values correspond to the
letters: $a = 0$, $b = 1$, $c = 2$, $d = 3$, $e = 4$, and so on; then the distance can be expressed by

$$e, \qquad c + c, \qquad d + b, \qquad e + a,$$

$$be, \qquad h - d, \qquad j - f, \qquad i \,/\, c.$$

By combining numbers and constants, the same distance may be
described also by

$$c^2, \qquad 2c, \qquad d + 1, \qquad e + 0,$$

$$4b, \qquad h - 3, \qquad d^2 - fb, \qquad 8 \,/\, c.$$

Since all of these expressions mean the same thing, any one of
them may be equated to any other. For example,

$$2 + 2 = 4, \qquad c + c = 2c, \qquad 4b = c^2, \qquad d^2 - fb = e.$$

Such expressions are more or less complicated than one another
but may each be particularly significant or useful in specific contexts. One such context is the analysis of location. Note, whereas
all the expressions above convey information about the length of
the rod, that is, the distance from one of its extremities to the
other, few of those expressions convey any information about the

location of the rod. At least the following expressions convey information about the location of the rod:

$$7 - 3, \qquad b - d, \qquad b - 3.$$

These expressions belong not only to arithmetic and algebra, but also to coordinate geometry. Along the X-axis, one extremity of the rod is aligned with the coordinate 3, the other extremity is aligned with the coordinate 7. To describe these coordinates with letters corresponding to the X-axis, they may be designated x_1 and x_2, respectively. Thus,

$$x_1 = 3, \qquad x_2 = 7,$$

and

$$x_2 - x_1 = 7 - 3 = 4.$$

The coordinates x_1 and x_2 designate the positions of the extremities of the rod along the X-axis.

We may speak of $x_2 - x_1$ as the length of the rod. But what if we write $x_1 - x_2$ instead? Then

$$x_1 - x_2 = 3 - 7$$

$$x_1 - x_2 = -4.$$

This result is peculiar. It raises the question: how can a length be negative? We sought to describe the distance between the two coordinates x_1 and x_2. There is only one quantity of units separating the two coordinates. Hence we should designate this quantity by a single number, or by expressions that represent a single number.

We can avoid the ambiguity by establishing the convention *not* to express any length by subtracting a coordinate having the greater numerical value from another having the smaller value. This convention suffices so long as we know the values of the coordinates. But what if we don't? Suppose the extremities of a rod are located at coordinates designated by x_a and x_b, but we do not know the numerical values or relative positions of these coordinates along the X-axis. How do we then express the length of the rod, the distance between the two coordinates? We cannot write simply

$$x_a - x_b \qquad \text{or} \qquad x_b - x_a,$$

because we do not know which of these expressions is incorrect, that is, which one has a negative value. We may avoid the problem by introducing another operation from ordinary algebra. By definition, the operation of obtaining an absolute value yields only positive numbers. Thus,

$$|3 - 7| = 4 \qquad \text{and} \qquad |7 - 3| = 4.$$

Hence we may well write either

$$|x_a - x_b| \qquad \text{or} \qquad |x_b - x_a|,$$

since

$$|x_a - x_b| = |x_b - x_a|,$$

that is, both of these expressions should now mean the same positive quantity. Thus, we might define the length between coordinates along a single reference axis as the absolute value of the two coordinates:

$$l = |x_a - x_b|,$$

regardless of whether the numerical value of x_a is greater or lesser than that of x_b. Nonetheless, if at least we know which of the two coordinates has the greater value, say, if x_b is greater than that x_a, then we may simply write

$$l = (x_b - x_a).$$

Up to this point, everything that has been said about the representation of the concept of length is part of the well-known and traditional approach to its mathematical analysis. For the most part, this traditional approach is an effective means of representing lengths symbolically. Nonetheless, ambiguities remain that now will be analyzed and clarified by means of nontraditional devices.

To begin with, note the curious asymmetry regarding the treatment of positive and negative numbers. We disallowed the notion of negative lengths while assuming it fair to conceive of all lengths as positive. The main reason was that (given one kind of measuring unit) there is only one distance between any two points, so this distance should be represented by a single number. Indeed, when measuring any object with a ruler we never obtain a negative number. But if there are no negative lengths, what do we mean by designating a length as positive?

Historically, the notion of positive numbers emerged only once the notion of negative numbers had been conceived. Previously, there was no need to ascribe the label "positive" to any numbers; they were just quantities. The concepts of positive and negative numbers were devised in response to practical contexts where some notion of opposition was involved. In particular, positive and negative numbers served to represent assets or debts, as well as movements, say, in one or another direction along a line. We could perhaps stretch such meanings to somehow relate to the concept of length, but in our ordinary use of this concept it does not involve any such notion of opposition. To be sure, we speak of distances traveled in opposite directions, but that is a matter that concerns motion, displacement. It is superfluous in the preliminary study of lengths. A distance is the same whether you measure it starting from one extremity or from the other. This impression is so pervasive that whenever we obtain different results for a length measured in opposite directions we assume that it must be merely the result of our carelessness.

Accordingly, it is superfluous to describe a length as "positive." Notice that this understanding is already at play in the ordinary way of writing lengths numerically. We write 3 and 7 instead of +3 or +7. It is an ancient habit, preceding the invention of negatives. Mathematicians now construe numerals without sign as simply a shorthand way of writing positives, but in actual practice we do not really bother to conceive of lengths as positive at all since there is no meaning to the concept of negative length. A length is just a quantity of units.

The expression "absolute value" is suggestive of the notion in question. Yet, traditionally, the mathematical operation of absolute value is conceived to result in positive numbers. Here we reach more asymmetries. Why does the operation called "absolute value" convert negative numbers into positives while leaving positives unaffected? Should not this operation more properly be called "positive value"? Why isn't there a converse single operation that yields negative values?

To solve these ambiguities, we may retain the notion that lengths should be described by "absolute values," while discarding the idea that absolute values are positive values. Thus, we may now operate with three kinds of numbers: positive, negative, and absolute values, that is, signless numbers. Hence we have

positive numbers: $^+1, ^+2, ^+3, ^+4, ^+5, \ldots$

negative numbers: $^-1, ^-2, ^-3, ^-4, ^-5, \ldots$

signless numbers: $0, 1, 2, 3, 4, 5, \ldots.$

Notice that the set of signless numbers includes the number 0, the one number at least that traditionally has been acknowledged to be neither positive nor negative. It is a signless number.

Signless numbers are simply a means of expressing the concept of quantity symbolically. They serve to convey that which numbers such as $^+2$ and $^-2$ have in common. Historically, the notion was occasionally distinguished from positive numbers, even if the distinction was not formulated symbolically. For instance, the first edition of the *Encyclopædia Britannica* noted that "though $+a$ and $-a$ are equal as to quantity, we do not suppose in algebra that $+a = -a$. . . ."[1] Signs served to represent qualities while signless numbers represented quantities. After the invention of negatives, mathematicians began to describe ordinary numbers as positive, yet this usage and its particular connotations never completely replaced the older notion of signless numbers. Whereas mathematicians advocated the identity between positives and signless numbers, their language often betrayed subtle distinctions.

Not only did signless numbers precede the invention of negatives but, even thereafter, they were occasionally distinguished overtly from positives. For example, John Farrar noted briefly that by contrast to the expressions $+4$, $+a$, -4, $-a$, "The quantity 4 or a may be considered independently of its sign."[2] Following Carnot, some writers considered "absolute numbers" as abstractions of quantities independent of any qualities or signs.[3] William Rowan Hamilton, in his *Elements of Quaternions*, briefly argued in a footnote that magnitudes, such as lengths, are best represented by numbers without signs: "This *number*, which we shall presently call the *tensor* ... may always be *equated*, in calculation, to a *positive scalar*: although it might perhaps more *properly* be said to be a *signless number*, as being derived solely from comparison of *lengths*, without any reference to *directions*."[4] Furthermore, the philosopher Bertrand Russell later also distinguished between signed and signless numbers: "we must, in the first place, clearly realize that numbers and magnitudes which have no sign are radically different from such as are positive. Confusion on this point is quite fatal to any just theory of signs."[5] But for the most part mathematicians and physicists routinely equated positive and signless numbers by considering the latter merely as a shorthand way of expressing the former. In discussing the "types" of numbers, writers and teachers still today do not distinguish between positive and signless numbers. Nevertheless, for our purposes it is better to distinguish between the two concepts. We may thus say not only that $^+4$ and $^-4$ are distinguished by their signs but also that they have the same thing in common: the magnitude that may be expressed by the signless number 4.

If later tonight you go to sleep and tomorrow somehow forget everything else in these pages, then at least please remember this: a distance of four is not the same as a displacement of four steps to the right. Just as it is not the same as a displacement of four steps to the left. Physically, four and positive four are not the same thing, at least not in the context of motion. If you agree with this point, then you might want to ponder or develop an arithmetic and algebra that do correspond to such empirical notions.

In light of the concept of signless numbers, we can return to the old problem of formulating a "science of quantity." Remember that to avoid the problems that stemmed from interpreting ambiguous mathematical expressions as quantities, mathematicians redefined the nature of mathematics, on the whole, so that it was no longer regarded as concerning essentially quantities. But the question remained: since traditional mathematics is not the science of quantity, then is there, nonetheless, or can there be, a science of quantity? Perhaps we can use the concept of signless numbers to formulate such a science. It would serve to describe and analyze physical relations among physical quantities. Specifically, we will now try to develop an arithmetic and algebra of quantity.

To begin, we may define the properties of signless numbers in relation to other numbers, as well as traditional operations.

Consider the operation of absolute value. Traditionally, it serves to convert negative and positive numbers into positive numbers. But rather than, again, granting such privileged status to positive numbers, we can redefine this operation. Given the three kinds of numbers, the revised operation of absolute values serves to convert any such number into a signless number. For example,

$$|4| = 4$$

$$|^+4| = 4$$

$$|^-4| = 4.$$

Only now does the expression "absolute value" correctly describe what this operation does.

With the new concept of signless numbers and the redefined operation of absolute value, we may continue to analyze the algebraic representation of lengths. To represent the position of points to the right or left of a reference mark on a line, we use positives and negatives, respectively. To describe a length, the quantity of units of a certain measure separating two points, we use absolute values. We may thus write

$$|^+7 - {}^+3| = |^+4| = 4$$

$$|^+3 - {}^+7| = |^-4| = 4$$

$$|^-7 - {}^-3| = |^-4| = 4$$

$$|^-3 - {}^-7| = |^+4| = 4.$$

Notice that we have retained the ordinary rules for the addition and subtraction of positives and negatives. Given the refined notions, the expressions

$$l = |x_a - x_b| \quad \text{and} \quad l = |x_b - x_a|$$

continue to be true in every case, whereas the expressions

$$l = (x_a - x_b) \quad \text{and} \quad l = (x_b - x_a)$$

cease to be true for any case, strictly speaking, since we now represent lengths only with signless numbers. Later, we shall find that the expressions consisting merely of the subtraction or addition of negative and positive numbers will have an exact use and meaning in the analysis of motion. But for now, consider the pair of absolute value expressions.

We may now ask, why do we require two operations to obtain lengths? Do we really need to first subtract one number from the other and then extract the absolute value? What we seek to represent is the difference between two coordinates, but ordinary arithmetical subtraction does not suffice to immediately and generally yield the single number of units separating the two coordinates. Subtraction yields different results depending on the order of the numbers. By contrast, the operation of addition produces results independent of the order of the numbers involved. We say that addition is commutative, meaning, for example, that

$$^+3 + {}^+7 = {}^+7 + {}^+3,$$

and generally that in *every* case, for any type of numerical values of *a* and *b*,

$$a + b = b + a.$$

We agree that sums are commutative, and we're all happy and satisfied about it. Meanwhile, we also feel quite comfortable with propositions such as

$$^+3 - {}^+7 \neq {}^+7 - {}^+3,$$

and generally that

$$a - b \neq b - a.$$

Why isn't the operation of subtraction commutative? There is nothing objectionable about such propositions of addition and subtraction; they are evidently true. Nonetheless, there is a formal peculiarity here. Addition and subtraction are fundamental operations of arithmetic, each being, it would seem, the opposite of the other. Yet, somehow, addition seems more basic, more elegant, owing partly to its commutative character. Wouldn't it be useful to have some sort of operation similar to subtraction but commutative? If so, we would then be able to express distances or lengths with a single operation as the "difference" between two coordinates, regardless of their order.

We may hence define a new operation. Let the separation of two coordinates, their "distinction," be given by the quantity of measurement units between the two. Symbolically, we may write

$$^+7 \sqcup {}^+3 = 4$$

$$^+3 \sqcup {}^+7 = 4$$

$$^-7 \sqcup {}^-3 = 4$$

$$^-3 \sqcup {}^-7 = 4.$$

Notice how the operation of distinction yields in a single step the results that had previously been obtained with two operations. To that extent, the results that we can express with the distinction operation can be expressed with the combination of subtraction and absolute value operations, if only we don't distinguish between positive and signless numbers. Nonetheless, this new operation is not just a shorthand way of writing expressions such as

$$|^+7 - {}^+3| = |^+4| = 4.$$

It does not require two operations performed in sequence to attain the result. It consists instead of a single individual assessment as immediate as performing an operation of addition. Any child can easily be taught that the distinction between 7 and 3 is 4, and is the same as that between 3 and 7; just as they learn that the addition of 3 and 4 is equal to the addition of 4 and 3. Algebraically, we may now define length simply by

$$l = x_a \sqcup x_b,$$

where

$$x_a \sqcup x_b = x_b \sqcup x_a.$$

We may likewise use the operation of distinction to express the "difference" between quantities of things. What is the difference between seven apples and three apples? Four. What is the difference between three apples and seven apples? Four. But notice that we must be careful with language, because, however appropriate the word "difference" may be presently, it is usually taken to designate the traditional operation of subtraction. It is for that reason that we speak of "distinction."

The introduction of signless numbers and the operation of distinction was rather simple. It serves to illustrate the ease with which elementary mathematics can be modified for the purpose of representing physical concepts, such as length, more directly and exactly. Of course, we haven't really "modified" traditional mathematics, we have just devised a new but similar set of rules, an artificial system. Anyhow, subtle and ambiguous questions begin to emerge soon after the new concepts are proposed. In particular, notice that we have not yet defined the meaning of ordinary arithmetical operations applied to signless numbers. We may begin this task as follows.

It is easy to imagine that the addition of signless numbers should proceed in the same way as the addition of positive numbers, so that, for example,

$$2 + 2 = 4.$$

Two quantities of units joined together constitute a length that is the sum of the quantities involved. However, what about the operation of subtraction? It is reasonable to expect that, say,

$$7 - 3 = 4,$$

but what are we to do with the expression

$$3 - 7 = ?$$

Historically, this is precisely the sort of problem that led people to invent negative numbers in the first place. But now that we have established here a distinction between positive and signless numbers, it seems inadequate to use ordinary negatives to express differences between signless numbers. The sum of signless numbers does not make positive numbers, so likewise why should their subtraction make negative numbers?

The invention of new types of numbers occurs precisely from attempts to provide solutions for such questions. Given that signless numbers represent quantities, someone could well take a bold step into the dark by inventing something called an antiquantity, for example, by writing

$$3 - 7 = \overline{4}.$$

But being interested in *practical* mathematics, we may ask: what the heck is an antiquantity? What does the number $\overline{4}$ represent?

Naturally, the inventor of antiquantities might want to defend the innocent, fledgling new numbers. Such a person might argue, for instance, that antiquantities necessarily exist as the solution to the problem posed, or that such numbers not only seem harmless but might well turn out to be quite useful someday, or that we should not stand in the way of what might actually be a major discovery simply because we don't yet understand it, and so on and so forth. After all, didn't people in the past express skepticism about and offer resistance against other types of once-mysterious-and-new numbers that are nowadays widely used, such as irrational numbers, negatives, and imaginaries? Of course, yes. Doubtless, such new sorts of seemingly impossible numbers have in the past served to advance science, partly by providing solutions to problems that otherwise would remain as obstacles preventing the analysis of greater puzzles. So fine, we can say, go ahead, study, cultivate, and play with your antiquantities, we hope some good will come out of it.

Nonetheless, there is something disappointing about this approach. Remember that at first we introduced the concepts of signless numbers and the operation of distinction as means to refine mathematical methods of representing the empirical notion of lengths. Although those concepts are not explicitly included in traditional mathematical texts, they constitute relations that are commonly at play in our ordinary calculational habits of thought. By comparison, the concept of an antiquantity seems extremely fanciful. First, we proposed simple improvisations akin to elementary mathematics essentially to convey more precisely and symmetrically concepts of physical significance. But all too quickly, we are suddenly inventing concepts that do not correspond straightforwardly or unambiguously to any sort of physical operation.

For our current purposes, the bottom line is that by itself the number $\overline{4}$ does not represent anything. We began this discussion simply as an analysis of the concept of length. We established a distinction between positive and signless numbers because we sought to reserve the designation "positive" as indicative of a certain directionality that the concept of quantity does not have, and, to attain a symmetry between negative and positive numbers that is otherwise lacking in traditional arithmetic. Instead of positives or negatives, signless numbers seemed appropriate for the description of quantities such as lengths. Hence we may seek the solution to the expression

$$3 - 7 = ?$$

by asking: exactly what do we mean by this expression? There is, to be sure, a degree of freedom in what meaning we choose to assign to an expression. But there is also a need for consistency. Since we defined the numbers 3 and 7 as quantities of units of length, the expression in question seems to be asking: what do you have left when you subtract seven units of length from three? Clearly, if we have three units of length, say three inches of tape measure, and we give away three, then we have nothing left. We can try to give away another unit of length but we can't because there is no more tape left. We can try to give away four more units, and we find that we can't do that either. We can promise people more tape, tell them we owe them tape, give them fancy receipts for the debt, or vouchers for future tape or for imaginary tape, or we can give them green apples without stems. But each of those transactions is a different story. No longer are we giving away tape. We cannot remove seven inches of tape from a strip that has only three left. The operation is impossible to completely carry out physically. Hence, *we wish to represent that mathematically.*

We can write

$$3 - 7 = \text{imp.},$$

or, say, we might write

$$3 - 7 = 0 \text{ inc.,}$$

meaning that we can subtract three of the seven from three, leaving zero, but that the rest of the subtraction cannot be completed, since there is nothing left. By this means, or by some other such alternative, we can express the physical understanding that one cannot withdraw a larger quantity from a smaller. The maximum quantity of units that can be removed from a length is the quantity of units that compose the length.

Again what we have here is simply an expression of a physical notion that is well known to common sense. Traditionally, the notion was discussed in introductory math texts. For example, Augustus De Morgan, in his textbook *On the Study and Difficulties of Mathematics* of 1831, endorsed the ordinary rules of operation of signs, but he established them on the proviso that if a be a number greater than b then the expression $+b - a$ is "incorrect," and so he distinguished between "real" and "mistaken" forms.[6] Likewise, William Kingdon Clifford later explained:

> if I write down the expression $3 - 7$, and then speak of it as meaning something, I shall be talking nonsense, because I shall have put together symbols the realities corresponding to which will not go together. To the question, what is the result when one number is taken from another, there is only an answer in the case where the second number is greater than the first.[7]

Despite statements such as these, mathematicians continued to allow subtractions of greater numbers from lesser numbers. Clifford explained that the "nonsense" should not be thrown away "as useless rubbish," but that it could be turned "into sense by giving a new meaning to the words or symbols." His solution to the dilemma was to interpret such subtractions in terms of steps along positive or negative directions. Nowadays this is commonly done and serves well for the study of motion, as it will for our purposes later on. Yet there it will concern operations with positive and negative numbers, while for now, in the study of the relations among lengths, we are concerned only with signless numbers. In writing, Clifford's general objective was to elucidate physical interpretations to justify the operations of ordinary mathematics. By contrast, our objective here is to devise new mathematics in accord with physical operations. While we could perhaps devise interpretations of 3 − 7 that yield a numerical solution, say, with antiquantities or otherwise, the point to emphasize is that we are essentially free to *decide* what sort of solution will be the result of such operations. It is a matter of how we choose to *define* the operation of subtraction of signless numbers.

We thus may decide, in accord with physical experience, that 3 − 7 is not an operation that can be carried out. Thus, subtractions of greater quantities from lesser quantities, are, in the words of Lazare Carnot, "un-executable operations." In a similar way, ordinary mathematics presupposes that division by zero is a meaningless operation.

Likewise, we may consider the operation of division applied to our signless numbers. Should we imagine that the division of a number by any other always yields a result? We may be comfortable with the operation

$$8 \div 2 = 4,$$

but what about

$$1 \div 3?$$

How do we divide a smaller quantity by a greater quantity? It might be that some physical things, such as certain very small material particles, are indivisible. If so, then the operation of division would not apply to them in a physical sense. Also, the actual space between two material particles might not in fact be infinitely divisible: there might only be a limited number of distinct positions that a particle can occupy in such a stretch. Moreover, even in everyday life, we may analyze the quantitative relations among objects that for whatever reason will not be individually divided. Accordingly, the division of signless numbers could be defined in a way that would place a limit on the extent to which a length may be subdivided.

However we define the operations of signless numbers, we next need to define the results of operations involving the mixture of signless and signed numbers. A simple numerical combination to begin with is multiplication. Our ordinary notions of multiplication harmonize well with the concept of signless numbers. For example, we say that "four apples times two is eight apples." This mode of expression is more natural than to say "four apples times two apples is eight apples." If we use a sign to represent a sort of thing, such as an apple, we can write

$$^+4 \times 2 = {}^+8.$$

Thus if we multiply a number of things the result is a number of things of the same sort. Multiplication by a signless number thus affects the quantity but not the quality or nature of the thing.

Likewise, if we need a distinct sign to represent things of a different sort, such as oranges, then we may write

$$^-4 \times 2 = {}^-8.$$

Now notice that in this representation the expression

$$^+4 \times {}^-4$$

is meaningless, for what does it mean to multiply four apples by four oranges? Here we could devise a system where minus times plus is an impossible operation. Nevertheless, what concerns us now is the establishment of rules relating signless numbers to signed numbers. Such rules may be useful both in contexts where the multiplication of signed numbers is meaningless and impossible, as well as in contexts in which it is meaningful.

As another context on which to ground the definition of operations, consider the analysis of motion. Here, the positive and negative numbers can represent motions in opposite directions. Again, the result of multiplying a signed number times a signless number may be defined to have the same sign as the signed number, meaning that the motion transpires in the same direction. For example, moving four feet to the right, times two, equals eight feet moved to the right.

The operation of addition may also be easily defined for the analysis of motion. For example, an object positioned at a coordinate $^+4$ can be said to move to the right by the addition of a signless number:

$$^+4 + 3 = {}^+7.$$

Here the result, $^+7$, represents the new position of the object.

Notice that these definitions of addition and multiplication are commutative; the result and its sign will be the same regardless of the order of the numbers. And addition seems as simple as multiplication because we have considered only positive numbers. The subject is not so simple once we consider subtraction as well as negative numbers. First, consider the addition of signless numbers to negative numbers. How should we define this operation? On the one hand, we can liken signless numbers to positive numbers, as if they were equivalent, and add them to negatives in the traditional way, for example,

$$^-4 + 3 = {}^-1.$$

But this is not really necessary; there are alternatives, because, on the other hand, we might instead establish that

$$^-4 + 3 = {}^-7.$$

Remember that we are free to define the meaning of the symbols as we see fit. And notice that only if we define the operation in the latter way do we arrive at the same property for addition that was valid for the operation of multiplication: that the combination of a signless number and a signed number yields *only* results that have *the same sign*. Otherwise, we might obtain, for example,

$$^-5 + 8 = {}^+3.$$

We might construe such usual results as "more natural," but only because we assume connotations of the traditional operation of addition. But once we introduce new elements, specifically, new numbers, into our arithmetic, we expand the traditional operations, and thus we essentially modify them. Thus we can define the addition of signless numbers to signed numbers as an increase in the quantity of such signed numbers. This definition might or might not be particularly useful in one context or another. But at least in the analysis of motion it introduces a perfect symmetry in the increase of motions in any directions.

Now consider the operations of subtraction and division. By analogy to the operation of addition, we can define subtraction of a signless number from a signed number as the decrease of the signed number. For example,

$$^-4 - 3 = {}^-1$$

$$^+4 - 3 = {}^+1.$$

In the analysis of motion we can assume that such decreases can move a body from one side of a coordinate axis to another, so that its position changes sign. For example,

$$^-4 - 5 = {}^+1$$

$$^+4 - 5 = {}^-1.$$

Notice again that the operations are thus perfectly symmetric. But note also that in this formulation "subtraction" is not the opposite of "addition," as we just established above that addition of signless numbers cannot change the sign of a number.

The operation of division can likewise be defined symmetrically when it involves signed and signless numbers. We may establish that

$$\frac{^-6}{2} = {}^-3 \quad \text{and} \quad \frac{^+6}{2} = {}^+3.$$

There is here nothing special about the positive direction as opposed to the negative direction. Such expressions are very well suited for the representation of motion. In particular, a signless denominator can represent time, while a numerator represents displacement in a specific direction, and the quotient represents a velocity in the same direction. There is no such exact representation in traditional arithmetic, where, for example, we would have

$$\frac{^-6}{2} = {}^-3 \quad \text{and} \quad \frac{6}{2} = 3.$$

Here it seems that, oddly, time has the same direction as motions to the right. But does time go toward the right? Of course not, it's just that the usual mathematical formalism does not distinguish among the positive terms. The distinction is made even more prominent when we describe other fundamental concepts of motion. By analyzing a motion in terms of its magnitude alone, regardless of direction, we *distinguish* speeds from velocities, and we *distinguish* distances from displacements. Or at least, we distinguish them physically and conceptually. However, traditional arithmetic involves no such exact distinctions, so that a motion considered *irrespective of direction*, for example,

$$\frac{\text{distance}}{\text{time}} = \frac{6}{2} = 3 = \text{speed},$$

is represented numerically in exactly the same way as a motion in a positive direction:

$$\frac{\text{displacement}}{\text{time}} = \frac{6}{2} = 3 = \text{velocity}.$$

The mathematical language does not capture the physical distinctions. However, the distinction is perfectly represented by the improvised numerical concepts

$$\left(\frac{6}{2} = 3 \right) \neq \left(\frac{^+6}{2} = {}^+3 \right).$$

This mode of arithmetical representation corresponds to the distinctions of vector theory, distinctions that otherwise lack adequate numerical equivalents. For example, vector theory does distinguish the concepts of motion:

$$\left(\frac{d}{t} = v \right) \neq \left(\frac{\mathbf{d}}{t} = \mathbf{v} \right).$$

To disregard such physical distinctions, as often done, can lead to ambiguities and mistakes.

Now, previously we defined an operation to convert any numbers into signless numbers; but how do we convert numbers into negatives and positives? The following operations are commonly employed:

> To make a positive number negative, we multiply it by $^-1$.
>
> To make a negative number positive, we multiply it by $^-1$,
>   or we may use the operation of the absolute value.

The last operation need not be considered since we have redefined it to produce signless numbers. Otherwise, to change the sign of a number, we multiply it by $^-1$. The procedure works for numbers as well as variables:

$$^+4 \times {}^-1 = {}^-4$$

$$^-4 \times {}^-1 = {}^+4$$

$$a \times {}^-1 = {}^-a.$$

But note that in the case of variables we do not know if the result is positive or negative, *unless* we know the sign of the number designated by the variable. We may ask, why does multiplication by minus one change signs, while multiplication by plus one does nothing? Why does the negative sign prevail over the positive sign when multiplied with it? Why does the multiplication of two negatives yield a positive? Is there physical meaning here? Yes, the negative sign is usually understood to mean that direction should be changed. But its association with multiplication is ambiguous and unnecessary. Moreover, the operation of changing the sign by multiplication does not by itself suffice to assign a definite sign to a variable. Accordingly, to assign a specific sign to any number we may establish the operations

$$^+|^+4| = {}^+|^-4| = {}^+|4| = {}^+4$$

$$^-|^+4| = {}^-|^-4| = {}^-|4| = {}^-4.$$

Likewise, for any variable

$$^+|a| = {}^+a \quad \text{and} \quad {}^-|a| = {}^-a,$$

where the signs mean that the value of the variable now *has* the noted sign. Now notice that here the superscripted signs are clearly not equivalent to the traditional plus and minus operations. For, if we write $+a$, in the usual interpretation, it does not tell us anything about whether a is positive or negative; and if we write $-a$ we change the sign of a but we yet have no knowledge of the new or previous value of a. Thus, if we establish distinctions between concepts by using signs, and systematically develop them, we can devise new mathematics. By distinguishing $+a$ from ^+a we conceive of two different operations. By distinguishing 4 from $^+4$ we conceived of different kinds of numbers. And so on.

Having defined several operations that relate signless numbers to signed numbers we might proceed to formulate an algebra. To be sure, we first need to establish arithmetical rules not only for the mixture of signed and signless numbers but also for mixtures of signed numbers. Such rules may be borrowed either from traditional algebra, or from any contrived system such as the one developed earlier, where minus times minus is minus. But if the algebra we seek to develop is to be a physical algebra then such rules should correspond to operations that have physical meaning. In any case, we may anticipate that an algebra based on the relations among three types of numbers will surely be more complicated than an algebra that involves only two sorts of numbers. This need not deter pursuit of the project, nevertheless, because in like manner there already exist several algebras that involve more than just positive and negative numbers, and yet they have fruitful applications.

Regardless, we will now set aside this important project, postpone it for some future work, in order to discuss instead the still incomplete foundations of the arithmetic and algebra of signless quantities alone. Previously, we noted that the arithmetic of signless numbers still entails ambiguities pertaining to the operation of subtraction. Such ambiguities would proliferate especially in an algebra of quantity, because, lacking knowledge of the numerical value of letters, it is more difficult to impose restrictions on the operation of subtraction. So how are we to formulate an algebra of quantity in which we do not generate new numbers, such as negatives or antiquantities?

We previously showed that it is possible to construct an algebra where there are no imaginary numbers; where the square root of a negative number is negative, and the square of a negative number is also negative, and so forth, all based on the rule that minus times minus is minus. Nevertheless, we still relied on the concept of negative numbers, the concept that individuals such as Maseres, Frend, and Hamilton identified as the fundamental source of the ambiguities of algebra. We also recognized that there are certain physical contexts where negative solutions are meaningless or impossible. Such contexts concern relations among magnitudes or quantities only, where it is meaningless to describe them as positive, negative, or having any other quality.

Throughout history, many algebraic analyses of physical problems have seemed mysterious. Algebra was criticized because its procedures were not as evident as those of geometry, wherein every step of a demonstration could be exhibited unambiguously. Nonetheless, physicists increasingly accepted algebra because it generally led to correct results, and they could routinely dismiss meaningless results. But if we desire a symbolic system that serves not only as a method of calculation but also as a mode of representation, then we should strive to develop an algebra in which every operation can be understood in physical terms. This is an important mathematical task: to develop precise descriptions of physical relations and processes.

If we are to have an algebra of quantity, then, we should expect that every operation should describe the relations and transformations of quantities. And having chosen signless numbers as the means for representing quantities, we should thus expect that any operations on these numbers should produce, again, only signless numbers. But this leads us back to the ancient problem that led originally to the invention of negative numbers. In order to construct an algebra, we need principles of transposition, that is, means by which to move quantities from one side of an equation to another. Traditionally, the simplest way to transpose a quantity, say, in an equation such as

$$a + b = c,$$

was to subtract the quantity from both sides. But this method of transposition often led to expressions such as

$$b - c = -a,$$

where the isolated negative sign seemed to indicate "a quantity less than zero." Arithmetically, this was the consequence of subtracting a greater number from a lesser. This method seemed necessary, however, in order to manipulate algebraic expressions, since the operation of subtraction was the only way of moving an added quantity, such as a or b, completely and by itself from one side of the equation to the other. Addition, multiplication, and division were inadequate for this purpose. Thus it might seem that subtraction is necessary and unavoidable as one of the fundamental operations of the science of quantity, an operation that entails the introduction of negative numbers and other concepts, suggesting, hence, that there can be no such system where the unrestricted manipulation of signless numbers leads only to other signless numbers. Nevertheless, this conclusion is mistaken.

There is an alternative. At our disposal we already have an operation that helps to formulate a consistent algebra of signless numbers. Consider again the operation of distinction. We saw earlier that distinction is a commutative operation:

$$a \sqcup b = b \sqcup a.$$

This property introduces into arithmetic a symmetry that is otherwise lacking owing to the operation of subtraction. For instance, it gives the operation of distinction a commutative identity element, something lacking in subtraction. That is, we may say that the commutative identity element of addition is zero because any number x plus zero, or zero plus any number x, results in that same number x. Likewise, the operation of distinction happens to have the same commutative identity element, zero. Subtraction has no such property.

What other properties does the operation of distinction have? Well, we readily notice that, unlike addition and multiplication, it is not associative:

$$(a \sqcup b) \sqcup c \neq a \sqcup (b \sqcup c).$$

For example,

$$(2 \sqcup 5) \sqcup 7 \neq 2 \sqcup (5 \sqcup 7)$$

$$3 \sqcup 7 \neq 2 \sqcup 2$$

$$4 \neq 0.$$

However, like addition, the operation of distinction has a distributive property with respect to multiplication:

$$a(b \sqcup c) = ab \sqcup ac.$$

For example,

$$2(5 \sqcup 7) = (2 \times 5) \sqcup (2 \times 7)$$

$$2(2) = 10 \sqcup 14$$

$$4 = 4.$$

Moreover, you can carry out numerical examples to see that distinction is also distributive with respect to division:

$$\frac{b \sqcup c}{a} = \frac{b}{a} \sqcup \frac{c}{a}.$$

These properties are immensely useful in the formulation of a general algebra.

Given the properties of distinction, consider now how we can manipulate quantities in an equation. First, as before, we attempt to transpose the quantities a and c in the equation

$$a + b = c.$$

By employing distinction instead of subtraction, we now have

$$a \sqcup a + b = c \sqcup a$$

$$0 + b = c \sqcup a$$

$$b = c \sqcup a$$

$$c \sqcup b = c \sqcup a \sqcup c$$

$$c \sqcup b = a.$$

Thus we can isolate any of the quantities a, b, c, without ever obtaining a quantity less than zero.

Now consider problems involving use of the four operations; addition, distinction, multiplication, and division. For example, let us transform the equation

$$a + \frac{b}{c} = d$$

in various ways to express the algebraic value of each letter. To begin with, to express the value of d, we also have

$$\frac{b}{c} + a = d,$$

according to the commutative property of addition. Likewise, to express a we have

$$a = \frac{b}{c} \sqcup d$$

$$a = d \sqcup \frac{b}{c}.$$

We can isolate b, by proceeding from this last equation, for example, as follows:

$$d \sqcup a = \frac{b}{c}$$

$$c(d \sqcup a) = \frac{b}{c} c,$$

to obtain the following expressions:

$$c(d \sqcup a) = b, \qquad c(a \sqcup d) = b,$$

$$cd \sqcup ca = b, \qquad ca \sqcup cd = b,$$

$$cd \sqcup ac = b, \qquad ca \sqcup dc = b,$$

$$dc \sqcup ac = b, \qquad ac \sqcup dc = b,$$

$$dc \sqcup ca = b, \qquad ac \sqcup cd = b,$$

$$(d \sqcup a)c = b, \qquad (a \sqcup d)c = b.$$

Thus we have the various possible ways to isolate the quantity b; various ways to correctly describe it in terms of the relations among the three other quantities. Finally, to express the quantity c, we have

$$c = \frac{b}{d \sqcup a}, \quad c = \frac{b}{a \sqcup d}.$$

By manipulating the symbolic expressions according to the established rules, we can ascertain any of the possible relations that will be true for *any* numerical quantities that satisfy, say, the first equation. For example, given again the expression

$$a + \frac{b}{c} = d,$$

we find arithmetically that it is true for the numbers

$$7 + \frac{6}{3} = 9.$$

Accordingly, every expression that is derived from the first will be valid for these numerical values. The symbolic expressions will describe purely arithmetical relations without generating nonarithmetical concepts. They will describe only and exclusively relations and transformations that are quantitatively possible. Hence, ladies and gentlemen, we have an algebra of quantity. Again, however, it is merely an example, a tentative and makeshift contrivance.

But what is the use of an algebra of quantity? Well, among other uses, it would be an adequate tool for describing the transformations of a closed physical system. For example, given a closed box containing twenty-three marbles, we may wish to describe the possible distributions and dynamics of the marbles. To describe the quantity of marbles located at one time in different sections of the box, we would employ the principle that there cannot be fewer than zero anywhere. And assuming that no more marbles can enter the box, we would expect that there cannot be more than all the marbles in the box, nor in any one section of it. If no marbles escape from the box, we could systematically describe possible distributions of marbles based on the initial quantity. The properties of the mathematics itself would facilitate representations of the possible intermediate states of the system as it transforms from one distribution to another. Thus, an algebra of quantity would be appropriate for analyzing systems where the total quantity of things is conserved, as a constant, and where hence there can nowhere be less than nothing nor anywhere more than everything.

And what about the operation of subtraction? We can introduce it, of course, if we encounter any need for it. But many problems do not require it. For example, "Tom has five apples and he gives three to Amy; how many apples does he have left?" Both subtraction and distinction give the correct result:

$$5 - 3 = 2$$

$$5 \sqcup 3 = 2.$$

Meanwhile, for the problem: "Tom has five apples and he gives eight to Amy; how many does he have left?" the operations give different results:

$$5 - 8 = -3$$

$$5 \sqcup 8 = 3.$$

We might think that here the operation of subtraction alone gives the correct result, but then again we are used to interpreting the result as satisfactory, after centuries of habituation. We interpret -3 as a debt. We do know, however, that the problem implies an operation that is physically impossible. Thus, we might employ the operation of subtraction precisely when we wish to describe physical impossibilities. In any case, we can, as usual, employ the concepts of negative and positive numbers, say in contexts where opposition is a subject of analysis; as with the analysis of directed motions, or the analysis of assets and debts. And there, subtraction can have a very definite meaning.

Returning to the algebra of quantity, note again that by no means is it a finished system. Although we can well use the four basic operations as the foundation upon which to build a more elaborate system, we can also refine or modify the foundations.

Consider one aspect that might be revised. Notice that three of the fundamental operations are commutative:

$$a + b = b + a$$

$$a \sqcup b = b \sqcup a$$

$$a \times b = b \times a.$$

But what about division? Division is not commutative:

$$a \div b \neq b \div a.$$

Might we not ascertain a similar operation, with clear physical meaning, that yet has the commutative property? We found great advantage in defining just such an operation as an alternative to subtraction; perhaps a symmetric sort of division may also be useful in describing relations among quantities.

By lacking the commutative property, subtraction and division also share a property that is easily illustrated in terms of physical considerations. Both of these operations presuppose that one quantity is to be taken *first*, in time, and then that some alteration be performed on it. For example, given four apples, divide them by eight. It matters, for the result, what quantity we *begin* with: eight or four. We can, to be sure, also give examples in which we formulate, say, multiplication in terms of one quantity given first. Yet we can also formulate multiplication as an operation concerning quantities that are given *simultaneously*. For example, consider the array

Here the three rows of units and the four columns of units constitute a total of twelve units. Hence we can write

$$3 \times 4 = 12.$$

The columns and rows, and, actually, all the quantities, appear simultaneously. Likewise, we can represent addition by means of a simultaneous arrangement of units:

such that

$$2 + 3 = 5.$$

We can also express distinction by particular arrangements

where we may write

$$3 \sqcup 4 = 1.$$

All three arithmetical expressions can be rewritten in a commutative manner without ceasing to describe the corresponding aggregates of quantities. But with division, a representation might correspond only to one of the possible divisions that can be carried out between two numbers. For example, given the arrangement

we can represent it by the equation

$$6 \div 3 = 2,$$

but the equation

$$3 \div 6 = \tfrac{1}{2}$$

does not represent the relations in question. In the array, three units are not divided into six parts. For that purpose, we might instead use an array such as

Now, considering the former arrangement, notice that there is really no sense in which the six quantities appear or are given *before* the three. We may read the array to mean that six quantities divided equally in three columns result in two quantities in each column. But we need not construe the six quantities as given first; we may simply establish the relationship that is simultaneously given among the quantities six and three. We may simply say that the relation between the quantities six and three is two. Or equally, the relation between three and six is two. So we can define an operation of "partition," such that

$$6|3 = 2, \qquad 3|6 = 2,$$

and, generally, that

$$a|b = c, \qquad b|a = c.$$

We can define this operation as the inverse of multiplication. "But wait," you might say, "isn't division already and really the inverse of multiplication?" So they say. Regardless, all we need for now is a guiding notion to help us establish values for the new operation. And we may establish such values on the basis of multiplication. For example,

$$2 \times 3 = 6: \qquad 2 = 6|3 = 3|6, \qquad 3 = 6|2 = 2|6;$$
$$3 \times 3 = 9: \qquad 3 = 9|3 = 3|9; \qquad 3 = 9|3 = 3|9;$$
$$4 \times 3 = 12: \qquad 4 = 12|3 = 3|12, \qquad 3 = 12|4 = 4|12;$$

and so forth. But we soon encounter ambiguities.

For example, if we define partition in terms of multiplication, then there yet remain many conceivable expressions that do not correspond to any multiplication. For example, the partition

$$a = 10|3$$

does not correspond to any multiplication

$$a \times 3 = 10$$

in which a is a whole number, such as 0, 1, 2, 3, 4. We might here proceed as in traditional division, namely, to invent a number defined by the operation:

$$10 \div 3 = \tfrac{10}{3}$$

and call any such numbers fractions. This will work for some purposes, but not for others. As noted earlier, depending on the nature of the objects or relationships described and analyzed by means of signless numbers, there are contexts wherein the concept of a fraction does not describe something physically possible. For example, ". . . just as negative numbers are sometimes non-sense, so are fractions. A polygon cannot have a fractional number of sides. A ball cannot be thrown $2\tfrac{1}{4}$ times. A surface has 2 dimensions, a solid 3; there is nothing in between."[8] In such cases where we cannot subdivide physical quantities, we can leave the operation of partition among such numbers undefined. Thus we may set this ambiguity aside, at least for the present purpose of formulating an elementary algebra of quantity.

But again we hit another difficulty. Given an equation of partition among two quantities, how do we move one quantity to the other side of the equation? If we establish that it is transposed by multiplying both sides of the equation by the number we wish to move, we find the result we hope for:

$$2 = 3|6$$

$$3 \times 2 = (3|6) \times 3$$

$$3 \times 2 = 6,$$

but this only works for *one* of the numbers:

$$2 = 3|6$$

$$6 \times 2 \neq (3|6) \times 6$$

$$6 \times 2 \neq 3.$$

And this difficulty becomes greater in algebra, where we no longer know even which one of two related values, a or b, can be transposed by multiplication:

$$c = a|b.$$

Despite such difficulties, we need not discard the operation of partition. It should not be surprising, after all, to find that a new operation does not follow the familiar and convenient properties of traditional operations. Perhaps in the difficulties raised by a new operation we can identify aspects of its arithmetical consequences that require attention. In the present case, notice that for an equation such as

$$2 = 3 | 6$$

it is *the smaller* of the related numbers, 3, not 6, that can be transposed by multiplication. This realization serves not only to formulate an arithmetical rule that applies to most combinations of numbers, but also to elucidate a principle on which to formulate a corresponding algebra. That is, we can formulate an algebra that takes into account the relative sizes of quantities.

We can describe relative sizes algebraically by stipulating that letters higher in the alphabet describe greater quantities. Thus, if c is greater than b, and b is greater than a, then the equation

$$a = b | c$$

implies that

$$a \times b = c.$$

Consequently, you might suddenly realize that the latter equation, too, implies already that c is the greater quantity.

Thus the algebra of quantity already incorporates ancient notions that are otherwise absent in traditional algebra. In particular, the product of two quantities multiplied the one by the other is greater than either of the quantities. You might object that this principle does not apply when any of the numbers multiplied is zero or one, and indeed it doesn't; but then again, to "multiply" by zero or one is just like not multiplying at all. Thus someone might argue that zero and one are not really "quantities," that is, that they are not plural entities. Hence, we can understand why some ancient civilizations, with good reasons, did not consider one or zero to be numbers. But fear not! We have excluded enough numbers in this conceptual exercise! We keep the special numbers 0 and 1, since they serve so well to describe the absence of quantity, and the fundamental unit of quantity, respectively. But we must be careful to note that they have unique properties with regard to certain operations.

Continuing, we should note now that in the expressions

$$a + b = c$$

$$b + a = c,$$

the quantity c must also be greater than a or b, unless either of the latter is zero. Also, in the expressions

$$c \sqcup a = b, \qquad c \sqcup b = a,$$

$$a \sqcup c = b, \qquad b \sqcup c = a,$$

we may accordingly establish that c is the greater quantity. To avoid inconsistencies between the results of addition and the results of multiplication, we may also write

$$a \times b = d$$

$$b \times a = d,$$

where d is also greater than a or b. And accordingly,

$$d|a = b, \qquad d|b = a,$$

$$a|d = b, \qquad b|d = a.$$

Given such conditions on the relative size of quantities, we may proceed to derive various relationships among quantities. For example, from the equation

$$a + b = c,$$

and given the partition values of a and b, it follows that

$$b|d + a|d = c$$

$$b|d + a|d = a + b.$$

Thus we may ascertain and exhibit general properties of the algebra.

But these last equations, what do they mean? They do express relationships among physical quantities, as you can figure out. Nonetheless, overall, notice that the way in which we established the operation of partition was more abstract than the way in which we established the operation of distinction. In the case of distinction we began from the soil of specific physical concepts, mainly the concept of length. For partition, however, we began from formal considerations about the algebraic properties of operations. Accordingly, partition remains a more ambiguous operation; it has not been sufficiently substantiated with empirical examples. Thus it is not as easy to understand or anticipate how this operation will behave in algebraic expressions, nor even in many arithmetical cases.

This newfound ambiguity should remind us to be careful in formulating the elements and rules of a novel mathematics. Once you begin establishing rules for the interrelation of symbols it is often tempting to devise various possible combinations as if it were only a game. This play of symbols, irrespective of the notions they represent, can easily become blind mathematics. Moreover, it may yield operations that are practically meaningless or impossible.

Even when elementary mathematical rules are designed to represent plain manipulations of things, they might still be used to construct symbolic statements that do not correspond to anything that can be exhibited palpably. Traditionally, in this sort of predicament, people proceed to formulate new interpretations that will give meaning to the otherwise meaningless expressions. So it was with "impossible" expressions. This approach can be a fruitful way of making sense of something that first appears useless. However, one might be driven by a trust in the general truth of symbolic expressions that can lead to problems. In particular, we may end up with a mathematics in which the various rules and their combinations are variedly justified in terms of mutually inconsistent analogies: considerations of quantity, space, motion, money, and so on. Throughout history, mathematicians realized that by adopting diverse and particular empirical explanations to justify special symbolic operations, mathematics acquired a semblance of arbitrariness and inconsistency. Thus they came to cast aside empirical explanations as mere illustrations and applications, and not as justifications for mathematical rules. To avoid this problem of defining rules on the basis of incompatible analogies, we should try to be carefully aware of the limited range of applications in which rules we have designed are appropriate and meaningful.

We conclude now the experimental portion of our discussion. We have devised the skeletal outlines of a few artificial numerical algebras. One system was characterized by the rule minus times minus makes minus. Another system introduced signless numbers alongside positives and negatives. Another was characterized by the requirement that only signless numbers be employed and all operations be commutative. We likewise suggested the possible formulation of several other systems: one where variables are manipulated according to their relative sizes; another where a greater quantity cannot be subtracted from a lesser; and yet another where the multiplication of different signs is meaningless, like multiplying apples with oranges. Such simple examples suffice to illustrate the subjects at hand. But caution, once manufactured, any such system can readily be divested of the notions that originally determined its rules. Thus it is important to remember that in each particular case we have adopted or stipulated rules owing to their significance. Our examples were tailored to correspond to notions about the symmetry of directions in space, the limits of the physical operation of subtraction, the distinction between distance and displacement, the absence of orientation in the notion of length, and so forth. It is mainly in connection to such contexts that such artificial systems of rules will be meaningful.

PHYSICAL MATHEMATICS?

The expression "physical mathematics" may seem to suggest a kind of paradox. How can mathematics, which is something intrinsically abstract, be physical? Of course, we're used to the notion of "mathematical physics," as we expect physics to be based upon and elaborated by all the resources applicable from mathematics. But in contradistinction, can we formulate and develop fields of mathematics that are dependent on physical knowledge?

The subjects that have been discussed in this book intersect a concern of some physicists throughout history. It is a basic question. How can we best employ mathematics to represent and analyze physical phenomena? If there existed only one mathematics, then the task of relating it to physics would consist essentially of application. That is, it would consist of finding ways to apply mathematics effectively to solve problems. But given that there exist multiple and diverse mathematics, the relation between physics and math extends beyond questions of application.

There is also a labor of selection. Scientists sometimes need to compare mathematical methods to identify which work best to solve specific kinds of problems. In each branch of physics, specialists may try to find and choose the mathematics most suitable or convenient to a particular subject. In this regard, the physicist and teacher August Föppl argued: "Herewith it should not be said that mechanics, lacking willpower of its own, must be bound to the usual representation-forms of pure mathematics. Rather, it must be allowed the freedom, if possible, to aid itself with the means of expression best suited for its aims."[9] Furthermore, since there exist multiple and diverse mathematics, physical mathematics should involve efforts to innovate. It should take advantage of the plasticity of symbolic algebra in order to formulate new or refined concepts and methods, designed in accord with physical experience.

If these goals are admitted, then physical mathematics would seek to provide appropriate representations and useful methods with which to analyze phenomena. Its basic task would consist of devising tools for describing things and relations symbolically.

To make new math tools we can devise new rules, sharper concepts, and special methods to better describe specific kinds of physical relations. By developing such innovations systematically, we can formulate new geometries, arithmetics, algebras, and so forth. For example, the mathematician and physicist Morris Kline explained:

> Mathematics does not have at its disposal special arithmetics to treat all of the situations in which ordinary arithmetic fails. For example, there is no arithmetic which tells us how volumes of gases combine. Each distinct combination must be analyzed on the basis of physical knowledge about the molecules involved. But there are situations which warrant the introduction of special arithmetical concepts and operations. If an arithmetic, that is, the concepts and operations, accurately describes physical events and permits prediction of future behavior, just as ordinary addition predicts the result of combining two herds of cattle, then it is worth creating.[10]

Yet students of science and physics are not taught how to devise new mathematics. Rarely are they even taught how the common mathematical methods acquired their present form. In portrayals of the history of math, the notion that mathematical principles are *discovered* is still nowadays employed much more often than any notions that some of those principles are created, decided, or contrived. Hence it is easy to get the impression that all principles are essentially discoveries.

One way to counter that impression is to study history, to see how mathematicians in the past have chosen, refined, and modified certain principles. Another way is to go ahead and devise new concepts and rules and develop algebras in which not all the traditional principles apply. This latter path can consist of a systematic game of arbitrary rules and symbols. It can be a self-serving end in itself. But it need not be so. It might also be steered by empirical notions.

As with symbolical algebra, the formulation of physical mathematics can be directed partly by notions of symmetry and simplicity. Preference for symmetry and simplicity stems from various roots. One motivation is aesthetic: the desire to elucidate graceful abstract systems. Another motivation is a kind of laziness: the inclination to get practical and general solutions while avoiding complications and work. Another motivation is the idea that symmetry and simplicity are qualities that pertain to truth.

And yet another motivation, related to these, is the desire to attain a symbolic language that matches the perceived order of things. In that sense, we might design new algebras endowed with appropriate simplicity for representing simple relations among specific things. We might try to develop an algebra that exhibits or focuses on symmetries that we observe in certain physical phenomena.

The converse is also valid. Physicists encounter plenty of phenomena and processes that are hardly simple. They encounter plenty of physical relationships that are not symmetric, say, where symmetries that might have been expected seem instead to be broken. Perhaps one might devise new algebras that capture such complexities.

Standing in the way of such possible approaches, there exists a practical complacency with traditional methods. We tend to value simplicity in our mathematics wherever it already exists, and we tend to accept as necessary any well-known complexities. A newly contrived math might likely be criticized for being simplistic wherever traditional math is complex, and it will be criticized for being complicated wherever traditional math is simple. Attempts to reform mathematics, or math education, are often met with resistance. Moreover, resistance tends to be greater especially wherever simple parts of traditional math seem to nicely describe aspects of physical phenomena that are apparently simple.

Many scientists hope that physical relationships can be represented best by the simplest mathematical models. But often that is hardly the case, and more complicated models are necessary. For centuries astronomers and philosophers imagined that the paths of the heavenly bodies were circular, as they deemed the circle to be the simplest geometrical figure. Unfortunately, such simple and elegant mathematical descriptions were mistaken. Heavenly bodies do not move in circles—none of them do. Subsequently, some people came to believe that heavenly bodies move in elliptical paths, but that too was a simplification. Thus, oftentimes the simplest mathematical accounts available do not exactly describe the phenomena or the physical structures. One physicist cautioned: "As far as the laws of mathematics refer to reality, they are not certain; and as far as they are certain, they do not refer to reality."[11]

Nevertheless, some inexact representations are very useful. But then we should try to ascertain how well such an account also describes actual physical structures, relationships, or interactions. Just because a formula works well to predict the outcomes of certain phenomena, that does not necessarily mean that it is a complete description of them. It might not even describe the underlying processes at all. With mathematical analysis we can systematically correlate the initial and final conditions of a physical problem. But that by itself does not necessarily involve representations of the intermediate physical processes. Hence it does not necessarily entail any profound physical understanding.

There is a habit, among some people, of interpreting the evolution of certain mathematical functions as literal descriptions of the invisible transformations of physical systems. One wellspring of this habit lay in teaching. After all, educators invest much time instructing students on cases where equations nicely match idealized physical processes. Far less time, if any, is invested in elucidating the limits of applicability of equations, cases in which the math does not match the physics. But by highlighting also the latter, perhaps students might more clearly develop a notion that the algebraic account in many cases is just a line sketch of the physical system.

Long ago, algebraists realized that they could find correct solutions to many problems despite using procedures involving steps that seemed unintelligible. The formulation of a given problem was often clearly meaningful, and the end result too, but some of the intermediate steps were not. At first, some individuals struggled to understand the intermediate steps. Some tried to devise alternative procedures that likewise would give the correct result without requiring blind steps. But eventually, most algebraists just accepted the idea that not every step in a computation needs to have a clear or physical meaning.

For example, George Boole claimed that "it is an unquestionable fact that the validity of a conclusion arrived at by any symbolical process of reasoning, does not depend upon our ability to interpret the formal results which have presented themselves in the different stages of the investigation."[12] He argued that when an algebraic calculus is applied to a given problem, producing a clearly interpretable and correct result at the end, then that serves to establish the validity of that specific application. Conversely, he did not argue that a physically incorrect result constitutes any evidence against math itself. It was usual among mathematicians, as it is today, to not impute to algebra itself any blame wherever it fails to give completely meaningful results in any particular application.

At the same time, however, Boole emphasized a prime condition of valid reasoning: that the laws for the combination of symbols should be fixed by the meaning assigned to such symbols. Accordingly, we might expect that to design an algebra that is physically meaningful, the rules of that algebra should be tailor-made to describe the kinds of physical things and relations under consideration.

Not only the rules, but also some of the mathematical objects or elements can be tailor-made. Thus we have spent some effort trying to illustrate how even new kinds of "numbers" can be invented. Many mathematicians would deny this notion that numbers themselves can be the product of creative thinking. But some have argued otherwise. For example, Richard Dedekind argued that at least some kinds of numbers are "free creations of the human mind."[13] And even if we don't agree on which numbers, among the kinds widely known, actually meet this characterization, we can nonetheless substantiate it by purposefully designing artificial kinds of numbers.

Thus we can devise new mathematical concepts and rules aimed to meet specific practical purposes.

Yet this pursuit is rarely carried out nowadays. Scarcely anyone designs new maths as methods of representation. Virtually nobody imputes to traditional algebraic methods any blame for instances where the symbolism generates unsatisfactory or bizarre results. Instead, when equations seem to lack meaning, or seem to signify things quite unintelligible, especially in modern physics, we often hear the following attitudes. Some teachers tell students that the problem just cannot be understood in terms of common sense. Some say that language, ordinary words, are not appropriate for describing the underlying physical relations. They sometimes also say that diagrams cannot exactly describe the physical system. Some also say that usual notions of cause and effect are just inadequate for understanding certain microphysical processes. Some even argue that logic itself is misleading or inadequate. However, virtually nobody says that maybe it is the algebra that is defective.

Sometimes standard algebraic methods generate seemingly impossible physical relations. This happens even in fundamental physics. But rather than imagine that standard algebraic methods are intrinsically inadequate when applied to certain kinds of problems, physicists generally trust algebra. They remember particularly the many cases where algebraic methods have produced surprisingly accurate and meaningful results. They recall instances where the math yielded surprising new discoveries. Thus the idea seems alien: that such a useful tool might be the faulty part in an investigation.

After all, standard algebraic methods give results that match a wealth of experimental results to very high degrees. Often the precision of mathematical predictions in physics is far greater than the precision obtained by methods of inquiry in other sciences. Fair enough; yet the results need not be what needs to be improved. What about the intermediate steps?

Some theoretical physicists occasionally have attempted to ascertain whether there are hidden aspects of physical systems that are not captured by current theories. Some others have claimed to derive mathematical proofs that such hidden variables are impossible. But such attempts are carried out within the confines of traditional mathematical methods. It is yet conceivable that new and perhaps more complicated algebras might serve to map physical structures and relations that are not captured by ordinary methods.

Suppose you have a simple mechanical gadget with which you can peel an orange. It may work well enough, and maybe you even use it to peel pears and pineapples as well. But that need not deter you from wondering whether some other contraption might also do the job. Maybe you can design a tool that would not only peel the orange but serve to exhibit something about its inner structure in the process. That might sound pointless if we're interested only in peels, or if we think that there's nothing inside. Maybe it's easier to accept apparent impossibilities and to suppose that we just live at a time in history when some limits of knowledge have been reached. The end of the road; reached already by some clever guys in the 1920s?

Most areas of contemporary theoretical physics are far removed from the kind of questions about immediate empirical quantification that have been discussed in the present book. One generally doesn't design new methods of mathematical representation. The project of devising artificial quantitative systems diverges from the dominant approaches in physics that build upon traditional symbolic methods. The complexities involved are many.

Contemporary theoretical physics focuses on the analysis of relations that are invariant amongst multiple systems. It often begins with the definition of mathematical spaces with certain global structural properties. To the extent that such mathematical structures are posited at the start of an investigation, physicists thus proceed by following ways in which pure mathematicians frame an inquiry. For instance, they define vectors as elements in an abstract space, determined by preestablished rules. This approach facilitates considerable mathematical power, even for carrying out computations, because it immediately allows the physicist to employ a wealth of theorems and methods that have been developed beforehand for the given kind of abstract space. Yet this approach disregards the possibility of beginning instead by improvising tailor-made rules or axioms that might better describe the kinds of physical things under study. Accordingly, the physicist Banesh Hoffmann cautioned that

> the person who studies vectors only as a branch of abstract algebra, regarding them solely as elements in a vector space, obtains the precision and elegance and power at a price. For they come to him too patly for him to realize their worth. He gains no inkling of the motives that led to the choice of just these axioms and ab-

stractions rather than others, he is apt not to realize how ugly are the seams that join such neat algebraic abstractions to their applications. . . .[14]

Decades earlier, the mathematician and physicist Henri Poincaré already had argued at length that the principles or axioms of geometry are essentially "conventions" that are freely chosen.[15]

We choose to employ particular abstractions because we believe that they suitably capture the important aspects of the subjects under investigation. For example, if we need to ascertain relationships only among the outlined shapes of material bodies we might use traditional diagrammatic geometry. Or if we need to ascertain relationships among distances or lengths, we might use coordinate algebra. Or if we need to ascertain relationships among motions in various directions, we might use vector algebra. And so on. Yet in every case, when compared to a given physical system under study, the mathematical structures stand as abstractions that just hopefully map well some key aspects of the system. When we try to extend the application of a specific method to cover additional aspects of the physical system, we may find shortcomings and difficulties. For example, imagine trying to use traditional diagrammatic geometry alone, involving only straight lines and circles, to represent colors. It would seem that we need some auxiliary mathematical tool. Likewise, when we try to extend the application of a usual mathematical method to a realm for which it was not designed, we might encounter difficulties. By highlighting such limitations, students of physics may better realize that sometimes different tools are needed for analyzing different problems.

The mathematics commonly used in physics remains a mixture of elements that vary widely in their physical significance, including some that appear meaningless. Maybe one reason why modern physics has been characterized often as involving aspects that are puzzling or unintelligible is because some of the mathematical rules and procedures it uses were not designed for its purposes. Oftentimes ordinary algebra is applied rather indiscriminately in contexts to which it is not exactly suited. And even though the results be correct, the procedures employed are hardly descriptive or direct.

Again, mathematical physics often tries to correlate physical objects and relations to abstract mathematical spaces. By contrast, the elementary kinds of questions discussed in the present book pertain to the representation of empirical quantities within a single system; they pertain to perceptible particulars. Mathematical physics nowadays, to the extent that it develops new methods, often deals with the question of modeling physical structures that are not perceptible. Nonetheless, in physics there is a growing field, located roughly midway between experiment and theory, which seeks to mathematically interconnect phenomena obtained experimentally. There one finds empirical data first and seeks to ascertain mathematical structures that capture such data. One seeks algebraic curves that accommodate a series of pointlike results, measurements. For such purposes, it might be useful not to restrict ourselves only to known mathematical structures but to devise appropriate new ones.

To the extent that physics analyzes empirical structures, it requires mathematical representations that describe such structures as closely as is possible or convenient. The experimental artificial algebras that are sketched in the previous sections were intended only to illustrate the plausibility of designing new mathematical systems. As they stand they remain insufficiently developed to be immediately applied to given physical theories. Nonetheless, we may ponder some plausible applications.

One virtue of an algebra based on the notion of length, for example, is that it would allow, like diagrammatic geometry, the systematic development of empirically meaningful relations. It could do so without invoking certain meaningless concepts such as that of negative lengths. Thus we may develop an algebra that systematically analyzes the possible relations and transformations of definite finite measurable lengths, as we encounter not only in basic experimental physics, but in design and architecture.

Separately, a noncommutative analytic geometry, such as the one formulated earlier, would enrich the study of material figures and their trajectories, by describing curves that are not captured by traditional analytic geometry.

Moreover, the project of formulating numerical systems where the rules of signs are perfectly symmetric would be appropriate in the empirical study of motion. The science of motion has been conceived often in two different ways. Some physicists construe it as the "pure" geometry of motion. Others construe it as the study of motion as it appears to observation. The prevalence of the former view allows this science to be construed commonly as derived from traditional mathematics. However, the latter view, literally and liberally construed, allows the possibility of formulating new mathematical concepts for the description of actual motions. This basic descriptive labor is at the root of physical mathematics.

We have considered ways in which new algebras can emerge when one requires deviations from some traditional rules. We saw that by abandoning the apparently universal validity of a basic rule, such as the commutative law of multiplication, new algebras were devised. Conversely, we also saw that we can devise new algebras by instead extending the applicability of that very rule, the commutative property, in designing artificial operations. Yet the possibilities for new algebras are still greater than those examples suggest. For instance, all of the examples in which we placed restrictions on the commutative property concerned only multiplication, but such a limitation is arbitrary. By trying to represent, say, ordinary notions of the displacement of bodies, we can even conceive of kinds of addition that are not commutative. Ordinarily, we require that vector addition is commutative, because we are usually interested, basically, in the initial and terminal locations of a body. But if we choose to emphasize instead the path of the body, its trajectory, then the addition of vectors in one sequence is not necessarily the same as the addition in another sequence. For example, to walk around a given barrier you may need to first walk to the right and then forward. This trajectory is not then the same as to try to walk these displacements in different order: first forward and then to the right. This latter combination of displacements might be impossible in the given situation, in which case we might say, by definition, that the one vector addition is not equal to the other. Accordingly, we may devise a vector algebra where addition is not commutative. This project, an algebra of transposition, is merely another hypothetical illustration of symbolical systems that we might devise.

Such plausible algebras might be long-term ventures. And such projects might not emerge from any urgent or obvious need as perceived by physicists who daily and successfully employ traditional methods. Are there deficiencies in current mathematical physics? If so, do we need to develop new mathematical methods? What kinds of methods would be preferable? Not everyone will give the same answers to such questions.

Nevertheless, at stake precisely is the question of cultivating, as a worthwhile goal, the pursuit of innovative symbolical systems as modes of representation. Therefore, we may try to innovate mathematical models that actually describe visible or invisible physical processes. The resulting mathematics would serve then not merely as an effective or interesting game of symbols, nor as an economical system of accounting, but as a language well suited for the description and analysis of physical systems.

Physical mathematics serves the aims of physics; it seeks to improve tools for representing and analyzing phenomena. We should actively try to develop concepts and methods that seek directly to describe physical relations and to serve specific needs.

For centuries, many philosophers and mathematicians have believed that physics should be based on mathematics, while mathematics should not be based on physics. To organize the various fields of knowledge according to their varying degrees of certainty, philosophers assumed that mathematics was more fundamental than physics, and thus that it would be logically wrong to ground mathematical rules on physical notions. Accordingly, despite the fruitfulness of physical considerations in the creation of new methods of analysis, many mathematicians attempted to expunge physical language, concepts, and analogies from the foundations of mathematics. Critical filters purified math of apparent soil and dirt. Hence physicists too came to construe mathematical truth as essentially independent of physical experience. Meanwhile, truth in physics was often deemed to be rooted in mathematical principles. Nature seemed to follow mathematical laws. Therefore, to understand natural phenomena, many natural philosophers advocated mathematics as the best means for understanding the order of things. Mathematics was said to be the language of nature.

But mathematicians proceeded to emancipate mathematics more and more from physical considerations, to farther extend it irrespective of correspondence with experience. The physical origins of basic mathematical rules were cast aside, obscured, and relegated to history; excised from the theory of mathematics. Thus it became a great mystery to understand why the results of such an allegedly pure and abstract system so often happen to correspond so well with physical relations.

Mathematics hence came to resemble less a physical language or science, and more an art or logic referring not to empirical relations but to abstract or transcendental principles. Some mathematicians even acknowledged that the rules of their symbolic algebras were arbitrary. As geometry and algebra were increasingly freed from the remnants of traditional restraints concerning what is possible and what is not, what is meaningful and what is not, they gave birth to a multitude of new abstract mathematics. Again, the new mathematics were applied occasionally to the description and analysis of nature, to various extents. Again, physicists did not require that propositions of such new mathematics be tailored and adapted to observable physical relations.

But not all mathematics need to be defined as irrespective of experience. As distinguished from abstract mathematics, physical mathematics can be based on patterns and notions derived from experience. Rules on the operation of symbols can be formulated specifically to describe physical operations and relations. Thus physical algebras can serve the needs of science. If so, it is yet conceivable that we may someday formulate a mathematics that could properly be characterized as the language of nature.

Physicists need not limit themselves to using common mathematical methods but should be free to improve them as necessary. We have seen examples of how this labor of innovation can be carried out. Once several tools are available for the representation and analysis of phenomena, there then follows a labor of selective use: to distinguish which of the different mathematical approaches serves best when applied to a specific sort of physical problem. Symbolic representations may thus be appreciated on the basis of their degree of correspondence to the empirical subject. Symbolic analyses may be judged on how exactly their results correspond to actual physical measurements, and also on the basis of how well the intervening mathematical steps describe the actual transformations of physical systems.

Even when one chooses to employ a well-established mode of mathematical analysis, one should at least try to distinguish which aspects of this mathematics correspond more closely to the physical subject at hand and which aspects do not. If this labor is neglected, then we are prone to arresting further physical inquiry by assuming that the current mathematical accounts already constitute complete and exact descriptions of yet unobserved physical structures and processes. It is thus easy to confuse a tool that serves well to predict the outcomes of certain kinds of physical interactions with a means of physical representation.

notes

Chapter 3: Much Ado About Less than Nothing

1. Nicholas Chuquet, *Triparty en la science des nombres* (1484); Michel Stifel, *Arithmetica Integra* (1544); for overviews, see Florian Cajori, *A History of Mathematical Notations*, vol. II (Chicago: Open Court, 1929), 126–131; Morris Kline, *Mathematical Thought from Ancient to Modern Times*, vol. I (1972, reissued, New York: Oxford University Press, 1990), 252–253; Kline, *Mathematics: The Loss of Certainty* (Oxford and New York: Oxford University Press, 1908), 115–116.

2. Hieronymi Cardani, *Ars Magna; Artis magnae, sive, De regvlis algebraicis, liber unus: qui & totius operis de arithmetica* (Norimbergae: Ioh. Petreium, 1545); trans. and ed. T. Richard Witmer as Girolamo Cardano, *The Great Art, or the Rules of Algebra* (Cambridge, Mass.: MIT Press, 1968).

3. Hieronymi Cardani, *De aliza regvla liber*, in *Opvs novvm de proportionibvs nvmerorvm, motvm, pondervm, sonorvm, aliarvmqve rervm mensurandarum . . . Praeterea. Artis magnae, sive, De regvlis algebraicis, liber vnvs, abstrvsissimvs & inexhaustus plane totius arithmeticae thesaurus* (Basileae: Henricpetrina, 1570).

4. R. C. H. Tanner, "The Alien Realm of the Minus: Deviatory Mathematics in Cardano's Writings," *Annals of Science* 37 (1980), 159–178.

5. F. Commandinus, *Euclidis Elementorum* (Pisa: Jacobus, 1572), f. 149.

6. Francisci Vietae, *Isagoge in artem analyticem* (1591); trans. T. Richard Witmer as François Viète, *The Analytic Art: Nine Studies in Algebra, Geometry and Trigonometry* (Kent, Ohio: Kent State University Press, 1983), 18–19.

7. Isaac Newton, *Universal Arithmetick*, trans. Mr. Ralphson, 2nd. ed., rev. Cunn (London, 1778), 3; reprinted in *The Mathematical Works of Isaac New-*

ton, vol. 2, ed. Derek T. Whiteside (New York and London: Johnson Reprint Corporation, 1967), 7; Leonhard Euler, *Vollständige Anleitung zur Algebra* (1770), trans. John Hewlett, 5th ed. (London: Longman, Orme, and Co., 1840; reprinted New York: Springer-Verlag, 1984), 1–5; Anonymous [Colin MacLaurin?], "Algebra," in *Encyclopædia Britannica; or, a Dictionary of Arts and Sciences, by a Society of Gentlemen in Scotland*, vol. I (Edinburgh: A. Bell and C. Macfarquhar, 1771), 80.

8. R. C. H. Tanner, "The Ordered Regiment of the Minus Sign: Off-beat Mathematics in Harriot's Manuscripts," *Annals of Science* 37 (1980), 127–158.

9. Kline, *Mathematics: The Loss of Certainty*, 115.

10. Carl B. Boyer, *History of Analytic Geometry* (New York: Scripta Mathematica, 1956), 86.

11. René Descartes, *Discours de la methode pour bien conduire sa raison, & chercher la verité dans les sciences. Plus La dioptriqve. Les meteores. Et La geometrie. Qui sont des essais de cete methode* (Leyde: I. Marie, 1637), pp. 372, 201, 380, 386; facsimile of the first edition, with translation and annotations by David Eugene Smith and Marcia L. Latham, *The Geometry of René Descartes* (New York: Dover, 1954), 159, 200, 175, 187.

12. Boyer, *History of Analytic Geometry*, 111.

13. John Wallis, *A Treatise of Algebra, both Historical and Practical. Shewing the original, progress, and advancement thereof, from time to time; and by what steps it hath attained to the heighth at which it now is* (London: R. Davis / J. Playford, 1685), 264; italics in the original.

14. Johannis Wallisii, *Arithmetica infinitorvm: sive nova methodus inquirendi in curvilineorum quadraturam, aliaq, difficiliora matheseos problemata* (Oxford: Leon. Lichfield, 1656), 74–75, 87.

15. Wallis, *Treatise of Algebra*, 264–272; italics in the original.

16. Newton, *Mathematical Works*, vol. 2, for example, 102–103.

17. Nicholas Saunderson, *The Elements of Algebra, in Ten Books*, vol. 1 (Cambridge: Cambridge University Press, 1740), 50.

18. Helena M. Pycior, *Symbols, Impossible Numbers, and Geometric Entanglements: British Algebra through the Commentaries on Newton's Universal Arithmetick* (Cambridge: Cambridge University Press, 1997), 266–269.

19. Colin MacLaurin, *A Treatise of Algebra, in Three Parts* (London: A. Millar and J. Nourse, 1748), 6.

20. *Ibid.*, 2–7.

21. Colin MacLaurin, "The Continuation of an Account of a Treatise on Fluxions, &c. Book II," *Philosophical Transactions of the Royal Society of London*, no. 469 (1742–43), 404–405.

22. Colin MacLaurin, *A Treatise of Fluxions in Two Books*, book 2 (Edinburgh: T. W. and T. Ruddimans, 1742), 577–578.

23. See, for example, Jean-Le-Rond d'Alembert, *Élémens de philosophie*

(1759); reprinted in *Oeuvres complètes de d'Alembert*, vol. 1, part 1 (Paris: A. Belin, 1821), 266.

24. Jean d'Alembert, "Négatif," *Encyclopédie*, vol. 11 (Paris: Briasson, 1765), 72, 655.

25. See also Jean d'Alembert, "Sur les Logarithmes des Quantités négatives," *Opuscules Mathématiques*, vol. 1 (Paris: David, 1761), 201.

26. d'Alembert, *Élémens de philosophie*, 267.

27. d'Alembert, "Négatif," 73.

28. Jean d'Alembert, "Sur les quantités négatives," *Opuscules Mathématiques*, vol. 8 (Paris: Jombert, 1780), 277.

29. J. d'Alembert and D. Diderot, "Impossible," *Encyclopédie*, vol. 8 (1765), 600.

30. Henrico Kuehnio, "Meditationes de quantitatibus imaginariis construendis et radicibus imaginariis exhibendis," *Novi Commentarii Academiae Scientiarum Imperialis Petropolitanae*, vol. III, for 1750 and 1751 (1753), 170–223.

31. Francis Maseres, *A Dissertation on the Use of the Negative Sign in Algebra: Containing a Demonstration of the Rules usually given concerning it; and shewing how Quadratic and Cubic equations may be explained without the Consideration of Negative Roots* (London: Samuel Richardson, 1758), ii–iii.

32. *Ibid.*, 2.

33. *Ibid.*, i.

34. *Ibid.*, 34.

35. Daviet de Foncenex, "Reflexions sur les quantités imaginaires," *Miscellanea Philosophico—Mathematica Societatis Privatae Taurinensis* 1 (1765), 113.

36. *Ibid.*, 123.

37. *Ibid.*, 113, 142.

38. Immanuel Kant, *Versuch den Begriff der negativen Größen in die Weltweisheit einzufüren* (Königsberg: Johann Jacob Kanter, 1763), 7–8; reissued in *Werke II: Vorkritische Schriften bis 1768/2* (Wiesbaden: Suhrkamp Verlag, 1960), 777–819; quotation on p. 785.

39. Leonhard Euler, *Vollständige Anleitung zur Algebra* (St. Petersburg: Kays. Acad. der Wissenschaften, 1770), secs. 142–143; emphasis in the original.

40. For a general account, see Charles Naux, *Histoire des Logarithmes de Neper a Euler*, vol. 2 (Paris: A. Blanchard, Libraire Scientifique et Technique, 1971), 154–189.

41. John Playfair, "On the Arithmetic of Impossible Quantities," read on 26 February 1778, *Philosophical Transactions of the Royal Society of London* 68, part I (1779), 318.

42. *Ibid.*, 320.

43. *Ibid.*, 321.

44. *Ibid.*, 322.

45. Sylvestre François Lacroix (1797), quoted in Boyer, *History of Analytic Geometry*, 212.

46. Auguste Comte, *Cours de Philosophie Positive*; trans. W. M. Gillespie as *The Philosophy of Mathematics* (New York: Harper & Brothers, 1851), 37.

47. J.-L. de La Grange, *Méchanique Analytique* (Paris: Veuve Desaint, 1788), vj.

48. Kline, *Mathematics*, 125.

49. Ellery W. Davis, *An Introduction to the Logic of Algebra*, 2nd ed. (New York: John Wiley & Sons, 1894), 46.

50. Jean Etienne Montucla, *Histoire des Mathématiques*, vol. 1, part 1, book 1 (Paris: Ch. Ant. Jombert, Imprimeur-Libraire du Roi, 1758, 2nd ed. Paris: H. Agasse, 1799), 33.

51. See, for example, George Berkeley, *The Analyst; or, a Discourse Addressed to an Infidel Mathematician* (Dublin: S. Fuller; London: J. Tonson, 1734).

52. George Berkeley, *A Defence of Free-Thinking in Mathematics. In answer to a pamphlet of Philalethes Cantabrigiensis, intituled, Geometry No Friend to Infidelity*, sec. XXI (Dublin: M. Rahmes for R. Gunne; London: J. Tonson, 1735), 7.

53. Berkeley, *The Analyst*, query 46.

54. Stendhal, *Vie de Henry Brulard* (written in the 1830s, first published in 1890; reissued, Paris: Le Divan, 1949), 375–376; italics in the original.

Chapter 4: Meaningful and Meaningless Expressions

1. Helena M. Pycior, "The Role of Sir William Rowan Hamilton in the Development of Modern Algebra," Ph.D. thesis, Cornell University, January 1976, 52. See also Augustus De Morgan, *A Budget of Paradoxes*, vol. 1 (1872, 2nd ed., ed. David Eugene Smith, Chicago and London: Open Court, 1915), 196–209.

2. William Frend, *The Principles of Algebra* (London: G. G. and J. Robinson, 1796), x–xi.

3. Ernest Nagel, "'Impossible Numbers': A Chapter in the History of Modern Logic," in *Studies in the History of Ideas*, ed. Department of Philosophy of Columbia University, vol. 3 (New York: Columbia University Press, 1935), 436; italics in the original.

4. Francis Maseres, "On the solution of certain cubick equations or equations of the Third Order, by Cardan's rules," in Frend's *Principles of Algebra*, 253; italics in the original.

5. Boyer, *History of Analytic Geometry*, 86. See, for example, Descartes, *Discourse on Method, Optics, Geometry, and Meteorology*, trans. Paul J. Olscamp (Indianapolis, New York, and Kansas City: Bobbs-Merrill Company, 1965), 198.

6. Boyer, *History of Analytic Geometry*, 111.

7. Caspar Wessel, "Om Directionens analytiske Betegning, et Forsøg, anvendt fornemmelig til plane og sphæriske Polygoners Opløsning," in *Nye Samling af det Kongelige Danske Videnskabernes Selskabs Skrifter*, vol. 5, Femte Del. Kjøbenhavn (1799), 469–518; trans. as *Essai sur la Répresentation Analytique de la Direction* (Copenhagen: Bianco Luno, Imprimeur de la Cour, 1897), 4–5.

8. Pycior, "Hamilton in the Development of Modern Algebra," 44.

9. *Ibid.*, 45.

10. Robert Woodhouse, "On the necessary Truth of certain Conclusions obtained by Means of imaginary Quantities," *Philosophical Transactions of the Royal Society of London* **91** (1801), 89–90.

11. *Ibid.*, 91–92.

12. *Ibid.*, 90.

13. *Ibid.*, 93, 119.

14. *Ibid.*, quotations are from pp. 95, 115, 108.

15. *Ibid.*, 119.

16. Pycior, "Hamilton in the Development of Modern Algebra," 64.

17. Robert Woodhouse, "On the Independence of the analytical and geometrical Methods of Investigation; and on the Advantages to be derived from their Separation," *Philosophical Transactions of the Royal Society of London* **92**, part 1 (1802), 86–87.

18. Lazare N. M. Carnot, *De la corrélation de figures de géométrie* (Paris: Duprat, Libraire pour les Mathématiques, 1801), 2, 24.

19. Charles Coulston Gillispie, *Lazare Carnot, Savant* (Princeton, N.J.: Princeton University Press, 1971), 122.

20. Lazare Carnot, *Géométrie de Position* (Paris: J. B. M. Duprat, Libraire pour les Mathématiques, 1803), 481.

21. Gillispie, *Lazare Carnot, Savant*, 126.

22. Carnot, *Géométrie de position*, xxxj, xij, xj.

23. Carnot, "Digression sur la nature des quantités dites négatives," in *Mémoire sur la rélation qui existe entre les distances respectives de cinq points quelconques pris dans l'espace; suivi d'un Essai sur la théorie des transversales* (Paris: Imprimeur—Libraire pour les Mathématiques, 1806), 108.

24. Gillispie, *Lazare Carnot, Savant*, 122, 140.

25. Lazare Carnot, *Réflexions sur la métaphysique du calcul infinitésimal*, 2nd ed. (Paris: Imprimeur—Libraire pour les Mathématiques, 1813), 201.

26. *Ibid.*, 227, 231, 217.

27. *Ibid.*, 225.

28. *Ibid.*, 218.

29. *Ibid.*, 188, 200.

30. Adrien Quentin Buée, "Mémoire sur les Quantités imaginaires," read 20 June 1805, *Philosophical Transactions of the Royal Society of London* **96**, pt. I (1806), 27; italics in the original.

31. *Ibid.*, 28; emphasis in the original.

32. *Ibid.*, 85.

33. R. Argand, *Essai sur une manière de représenter les quantités imaginaires dans les constructions géométriques* (1806, 2nd ed., Paris: Gauthier Villars, 1874).

34. *Ibid.*, 12; italics in the original.

35. *Ibid.*, 13.

36. For example, by François-Joseph Servois, "Lettre de M. Servois," *Annales de Mathématiques pures et appliquées, J. D. de Gergonne* 4 (1813), 84.

37. R. Argand, "Réflexions sur la nouvelle théorie des imaginaires, suivies d'une application à la démostration d'un théorème d'Analyse," *Annales de Mathématiques pures et appliquées, J. D. de Gergonne* 5 (1814–15).

38. Argand, *Manière de représenter les quantités imaginaires*, 60.

39. "Art. II. Memoire sur les Quantités Imaginaires. Par M. Buëe," *The Edinburgh Review* 12 (1808), 306–318; quotation on p. 310.

40. Sylvestre François Lacroix, *Élémens d'algèbre a lusage de l'École Centrale des Quatre-Nations*, 14th ed. (Paris: Bachelier, 1825), 88–110, 359–360.

41. Thomas Simpson, *A Treatise of Algebra: wherein the principles are demonstrated and applied in many useful and interesting inquiries, and in the resolution of a great variety of problems of different kinds*, 1st American ed. from 8th English ed. (Philadelphia: Mathew Carey / T. & G. Palmer, 1809), 24.

42. *Ibid.*, 25.

43. Jeremiah Day, *An Introduction to Algebra, being the First Part of a Course of Mathematics, adapted to the Method of instruction in American Colleges* (1814, 36th ed., New Haven, Conn.: Durrie & Peck; New York: Collins, Keese & Co., 1839), 16; italics in the original.

44. John Farrar, *An Introduction to the Elements of Algebra, designed for the use of those who are acquainted only with the first principles of Arithmetic. Selected from the Algebra of Euler*, 2nd ed. (Boston: Hilliard and Metcalf, 1821), 9.

45. Day, *Introduction to Algebra*, 324–325.

46. Augustus De Morgan, "On the Relative Signs of Coordinates," *Cambridge Philosophical Transactions* (1836); "On Leonardo da Vinci's Use of + and –," *Cambridge Philosophical Transactions* (1842); "On the mode of using the Signs + and – in Plane Trigonometry," *Cambridge and Dublin Mathematical Journal* (May 1851); "On the Signs + and – in geometry and on the Interpretation of the Equation of a Curve," *Cambridge and Dublin Mathematical Journal* (November 1852); "On the Early History of the Signs + and –," *Cambridge Philosophical Transactions* (1864).

47. Sophia Elizabeth De Morgan, *Memoir of Augustus De Morgan* (London: Longmans, Green and Co., 1882), 19.

48. Augustus De Morgan, *On the Study and Difficulties of Mathematics* (1831, reissued, Chicago and London: Open Court, 1902), vi, and chap. 9.

49. *Ibid.*, 71.

50. *Ibid.*, 72; see also Jean-Marie-Constant Duhamel, *Des Méthodes dans les*

Sciences de Raisonnement, pt. 2, 3rd ed. (Paris: Gauthier-Villars, 1896), chap. XIX.

51. De Morgan, "On the Negative Sign, etc.," in *Study and Difficulties*, 104.

52. De Morgan, *Study and Difficulties*, 105–109.

53. Augustus De Morgan, *Trigonometry and Double Algebra* (London: Taylor, Walter, and Maber, 1849), 113, alluding to the memoir in *The Edinburgh Review* 12 (1808), 308.

54. John Playfair, "Dissertation Third; Exhibiting a General View of the Progress of Mathematical and Physical Science since the Revival of Letters in Europe," *Dissertations on the History of Metaphysical and Ethical and of Mathematical and Physical Science, Encyclopædia Britannica*, 7th ed. (Edinburgh: Adam and Charles Black, 1835 & 1842), 443; italics in the original.

55. John Leslie, "Dissertation Fourth; Exhibiting a General View of the Progress of Mathematical and Physical Science, chiefly during the Eighteenth Century," *Encyclopædia Britannica*, 592–593.

56. John Warren, *A Treatise on the Geometrical Representation of the Square Roots of Negative Quantities* (Cambridge: J. Smith for Cambridge University Press, 1828), 1; italics in the original.

57. C. V. Mourey, *La vrai Théorie des quantités négatives et des quantités prétendues imaginaires* (Paris: Bachelier, 1828).

58. Carolo Friderico Gauss, "Theoria residuorum biquadraticorum. Commentatio secunda," *Commentationes Societatis Regiae Scientiarum Gottingensis* 7 (1832); reprinted in Carl Friedrich Gauss, *Werke*, vol. 2 (Göttingen: Königlichen Gesellschaft der Wissenschaften zu Göttingen, 1863), 95–148.

59. Carl Friedrich Gauss, untitled discussion in *Göttingische gelehrte Anzeigen* (23 April 1831); reissued in *Werke*, 169–178; quotation on p. 175.

60. See, for example, Comte's *Cours de Philosophie Positive*, 81.

61. John Bonnycastle, *A Treatise on Algebra, in Practice and Theory*, vol. II, 2nd ed. (London: J. Nunn; Longman and Co.; Cadell and Davies; John Richardson; Baldwin, Cradock and Joy; Sherwood and Co.; Ogle, Duncan and Co.; G. and W. B. Whittaker; John Robinson; Simpkin and Marshall, 1820), 4.

62. George Peacock, "Report on the Recent Progress and Present State of Certain Branches of Analysis," *Report of the Third Meeting of the British Association for the Advancement of Science* (1833), 185–352.

63. Peacock, "Report on the Recent Progress," 198–199. See also George Peacock, *A Treatise on Algebra* (Cambridge: Cambridge University Press, 1830), 104; *A Treatise on Algebra*, vol. 2, *On Symbolical Algebra and Its Applications to the Geometry of Position* (Cambridge: Cambridge University Press, 1845), 59, 448–449.

64. De Morgan, *Trigonometry and Double Algebra*, 92.

65. *Ibid.*, 101; italics in the original.

66. E. T. Bell, *Men of Mathematics* (1937, reissued, New York: Simon and Schuster, 1962), 354–355.

67. De Morgan, *Trigonometry and Double Algebra*, 94.

68. Arnold Dresden, *An Invitation to Mathematics* (New York: Henry Holt and Company, 1936), 35, 85.

69. *Ibid.*, 83, 86.

70. Edna E. Kramer, *The Main Stream of Mathematics* (New York: Oxford University Press, 1951), 113.

71. Bonnycastle, *Treatise on Algebra*, 4.

72. S.-F. Lacroix, *Essais sur l'enseignement en Général et sur celui des Mathématiques en Particulier* (Paris: Courcier, Imprimeur—Libraire pour les Mathématiques, 1805), 284.

73. E. T. Bell, *The Development of Mathematics* (New York and London: McGraw-Hill, 1940), 159.

74. Paul J. Nahin, *An Imaginary Tale: The Story of $\sqrt{-1}$* (Princeton, N.J.: Princeton University Press, 1998), 13.

75. Among many others, for example, see Cardani, *Ars Magna*; Euler, *Vollständige Anleitung zur Algebra*.

76. Ernest Nagel, " 'Impossible Numbers'," 435–436; italics in the original.

77. Nahin, *Imaginary Tale*, 37.

78. A similar example is found in Simpson's *Treatise of Algebra*, 27.

79. John Playfair, "On the Arithmetic of Impossible Quantities," *Philosophical Transactions of the Royal Society of London* 68 (1779), 318–319.

Chapter 5: Make Radically New Mathematics

1. Ludwig Schlesinger, "Über Gauss' Arbeiten zur Funktionentheorie," in Carl Friedrich Gauss, *Werke*, vol. 10, pt. 2 (Göttingen: Gesellschaft der Wissenschaften zu Göttingen / Julius Springer, 1922–1933), 55–57.

2. Felix Klein, *Vorlesungen uber die Entwickelung der Mathematik im 19 Jahrhundert*, pt. I (Berlin: Springer-Verlag, 1928); trans. M. Ackerman in *Development of Mathematics in the 19th Century*, vol. IX, *Lie Groups: History, Frontiers and Applications* (Brookline, Mass.: Math Sci Press, 1979), 110, 76.

3. Gottfried Wilhelm Leibniz, letter to Christiaan Huygens, 8 September 1679, in *Christiaani Hugenii aliorumque seculi XVII. virorum celebrium exercitationes mathematicae et philosophicae*, edited by Uylenbroek (1833), fasc. I, p. 10.

4. H. Halberstam and R. E. Ingram, introduction to *The Mathematical Papers of Sir William Rowan Hamilton*, vol. 3 (Cambridge: Cambridge University Press, 1967), xiv.

5. W. R. Hamilton to John T. Graves, 20 October 1828, in R. P. Graves, *Life of William Rowan Hamilton*, vol. 1 (Dublin: Hodges, Figgis, & Co., 1882), 304.

6. W. R. Hamilton, "Theory of Conjugate Functions, or Algebraic Couples; with a Preliminary and Elementary Essay on Algebra as the Science of Pure

Time," *Transactions of the Royal Irish Academy* 17 (1837), 293–422; reprinted in *Mathematical Papers*, 4.

7. Hamilton, "Theory of Conjugate Functions," 4; italics and emphasis in the original.

8. C. C. MacDuffee, "Algebra's Debt to Hamilton," in *A Collection of Papers in Memory of Sir William Rowan Hamilton* (New York: Scripta Mathematica, 1945), 26–27.

9. Hamilton, "Theory of Conjugate Functions," 5; see also Thomas L. Hankins, *Sir William Rowan Hamilton* (Baltimore: Johns Hopkins University Press, 1980), 250.

10. Hamilton, "Theory of Conjugate Functions"; Hankins, *Sir William Rowan Hamilton*, 254.

11. W. R. Hamilton, Abstract to "Account of Theory of Systems of Rays," presented to the Royal Irish Academy on 23 April 1827, reprinted in Graves, *Life of William Rowan Hamilton*, 228–229.

12. Immanuel Kant, *Kritik der reinen Vernuft* (1781); trans. and ed. Paul Guyer and Allen W. Wood as *Critique of Pure Reason* (Cambridge: Cambridge University Press, 1998), 144, 633–636. Also, Immanuel Kant, *Prolegomena zu einer jeden künftigen Metaphysik die als Wissenschaft wird auftreten können* (1783), §10, trans. Paul Carus and James W. Ellington as *Prolegomena to any Future Metaphysics that will be Able to Come Forward as Science*, 2nd ed. (Indianapolis and Cambridge, England: Hackett Publishing Company, 1977), 25.

13. Hamilton, "Theory of Conjugate Functions," quotation on p. 96.

14. *Ibid.*, 5–6.

15. See letter from Hamilton to John T. Graves, 17 October 1843, *Philosophical Magazine* 25 (1844), 489–495; reprinted in *Mathematical Papers*, 106–110.

16. Michael Crowe, *A History of Vector Analysis* (Notre Dame, Ind.: University of Notre Dame Press, 1967; reissued New York: Dover Publications, 1994), 25.

17. W. R. Hamilton, "Account of a Theory of Systems of Rays" (1827); in Graves, *Life of William Rowan Hamilton*, vol 1 229.

18. W. R. Hamilton, Quaternions, notebook 24.5, entry for 16 October 1843, published in *Mathematical Papers*, 103; Andrew Pickering, *The Mangle of Practice: Time, Agency, and Science* (Chicago and London: University of Chicago Press, 1995), 126–138.

19. For an explanation, see Pickering, *Mangle of Practice*, 133–136.

20. W. R. Hamilton, notebook 24.5, entry for 16 October 1843, published in *Mathematical Papers*, 103–105; also, "On a New Species of Imaginary Quantities Connected with the Theory of Quaternions," *Proceedings of the Royal Irish Academy* 2 (1844), 424–434; reprinted in *Mathematical Papers*, 111–116.

21. Hermann Grassmann, foreword to *Die lineale Ausdehnungslehre, ein neuer Zweig der Mathematik* (Leipzig: Otto Wigand, 1844), trans. Lloyd C.

Kannenberg as *A New Branch of Mathematics* (Chicago and LaSalle, Ill.: Open Court, 1995), 9.

22. A.-L. Cauchy, "Mémoire sur la théorie des équivalences algébriques," *Exercises d'Analyse* **4** (1847), 94.

23. W. R. Hamilton, "On Quaternions; or on a New System of Imaginaries in Algebra," *Philosophical Magazine* **29** (1846), 26–31; reprinted in *Mathematical Papers*, quotation on p. 236.

24. George Boole, *An Investigation of the Laws of Thought, on which are founded the Mathematical Theories of Logic and Probabilities* (London: Walton and Maberly; Cambridge: Macmillan and Co., 1854), 37–38, 48.

25. Crowe, *History of Vector Analysis*, 132.

26. James Clerk Maxwell, "Address to the Mathematical and Physical Sections of the British Association," *British Association for the Advancement of Science Report* **40** (1870); reprinted in W. D. Niven, ed., *The Scientific Papers of James Clerk Maxwell*, vol. 2 (Cambridge: Cambridge University Press, 1890, reissued New York: Dover Publications, 1965), quotation on p. 218.

27. J. C. Maxwell, "On the Mathematical Classification of Physical Quantities," *Proceedings of the London Mathematical Society* **3** (1871), 224–232; reprinted in Niven, *Scientific Papers of Maxwell*, vol. 2, quotation on p. 259.

28. J. C. Maxwell, letter to P. G. Tait, 9 October 1872, published in Cargill Gilston Knott, *Life and Scientific Work of Peter Guthrie Tait* (Cambridge: Cambridge University Press, 1911), 151.

29. Maxwell, "Mathematical Classification of Physical Quantities."

30. J. C. Maxwell, letter to P. G. Tait, 7 September 1878, published in Knott, *Life and Scientific Work of Peter Guthrie Tait*, 151–152.

31. Graves, *Life of William Rowan Hamilton*, 445.

32. Maxwell, "Mathematical Classification of Physical Quantities," quotation on p. 259.

33. P. G. Tait, "Hamilton," *Encyclopaedia Britannica* (1880); reprinted in P. G. Tait, *Scientific Papers*, vol. 2 (Cambridge: Cambridge University Press, 1900), doc. CXXVIII, quotation on p. 443.

34. *Ibid.*

35. P. G. Tait, "On the Intrinsic Nature of the Quaternion Method," *Proceedings of the Royal Society of Edinburgh* **20** (2 July 1894), 276–284; reprinted in Tait, *Scientific Papers*, vol. 2, doc. CXVI, 393; Tait, "Address to Section A of the British Association," *British Association Report*, Edinburgh, 3 August 1871; reprinted in Tait, *Scientific Papers*, vol. 1, doc. XXIII, 164.

36. P. G. Tait, "On the Rotation of a Rigid Body about a Fixed Point," read on 21 December 1868, *Transactions of the Royal Society of Edinburgh* **25** (1868), 261–303; reprinted in Tait, *Scientific Papers*, vol. 1, doc. XV, quotation on p. 86.

37. P. G. Tait, "On the Importance of Quaternions in Physics" *Philosophical*

Magazine **29** (1890), 84–97; in Tait, *Scientific Papers*, vol. 2, doc. XCVII, p. 299.

38. Alexander McAulay, *Utility of Quaternions in Physics* (London and New York: Macmillan and Co., 1893), 2; Alexander Macfarlane, "Review of *Utility of Quaternions in Physics*," *Physical Review* **1** (1893), 389.

39. Lewis Carroll, *Through the Looking-Glass: and What Alice Found There* (London: Richard Clay; New York: Macmillan, 1872), chap. 5.

40. Horatio N. Robinson, *New Elementary Algebra; Containing the Rudiments of the Science, for Schools and Academies* (New York: Ivison, Phinney, Blakeman & Co., 1864), 34–41; italics in the original.

41. Louis Pierre Marie Bourdon, *Éléments d'Algèbre, ouvrage adopté par l'université*, 14th ed., rev. M. E. Prouhet (Paris: Gauthier-Villars, 1873), 73–76.

42. Wm. Cain, *Symbolic Algebra, or The Algebra of Algebraic Numbers, Together with Critical Notes on the Methods of Reasoning Employed in Geometry* (New York: D. Van Nostrand, 1884), iii–iv, 50–56.

43. L. Kronecker, "Über den Zahlbegriff," *Crelle, Journal für die reine und angewandte Mathematik* **101** (1887), 337–355; also in *Philosophische Aufsätze, Eduard Zeller zu seinem fünfzigjährigen Doctor-Jubiläum gewidmet* (Leipzig, 1887), 261–274; reissued in *Leopold Kronecker's Werke*, edited by K. Hensel, vol. 3, pt. 1 (Leipzig: B. G. Teubner, 1899), 251–274; see sec. 5, 260–261.

44. Richard Dedekind, *Was sind und was sollen Zahlen* (1887, 2nd ed., Braunschweig: Friedrich Vieweg und Sohn, 1893); trans. Wooster Woodruff Beman, *Essays on the Theory of Numbers*, I. *Continuity and Irrational Numbers*, II. *The Nature and Meaning of Numbers* (New York: Dover Publications, 1963), 34–35.

45. Oliver Heaviside, "The Elements of Vectorial Algebra and Analysis," in *Electromagnetic Theory*, vol. 1 (London: The Electrician, 1893; reprinted New York: Chelsea Publishing Company, 1971), 137.

46. Heaviside, *Electromagnetic Theory*, vol. 3 (London: E. Benn, 1925), 135.

47. Heaviside, "Vectors versus quaternions," *Nature* **47** (1893), 533–534.

48. J. W. Gibbs, letter to Victor Schlegel, 1 August 1888, published by Lynde Phelps Wheeler, in *Josiah Willard Gibbs: The History of a Great Mind* (New Haven, Conn.: Yale University Press, 1962), 107.

49. For example, J. W. Gibbs, "On the Determination of Elliptic Orbits from Three Complete Observations," in *Memoirs of the National Academy of Sciences* **4**, pt. II (1889), 79–104; reprinted in *The Scientific Papers of J. Willard Gibbs*, vol. 2 (London, New York, and Bombay: Longmans, Green and Co., 1906; reprinted New York: Dover Publications, 1961).

50. Heaviside, *Electromagnetic Theory*, vol. 3, 137.

51. Oliver Heaviside, "On the Forces, Stresses, and Fluxes of Energy in the Electromagnetic Field," presented to the Royal Society, abstract in *Proceedings* **50** (1891), 126–129; *Transactions, A* **183**, (1892), 423–480; reprinted in

Heaviside, *Electrical Papers*, vol. 2 (New York and London: Macmillan and Co., 1894), quotation on p. 529.

52. Heaviside, *Electromagnetic Theory*, vol. 1, quotation in the Preface (p. ii); see also p. 305.

53. Cargill Gilston Knott, "Recent Innovations in Vector Theory," *Proceedings of the Royal Society of Edinburgh* 19 (1893), 221–222.

54. Crowe, *History of Vector Analysis*, 196.

55. Heaviside, "The Elements of Vectorial Algebra and Analysis," in *Electromagnetic Theory*, vol. 1, quotation on p. 133.

56. *Ibid.*

57. *Ibid.*, 134.

58. J. W. Gibbs, "Quaternions and the Algebra of Vectors," *Nature* 47 (1893), 463–464; reprinted in *Scientific Papers of J. Willard Gibbs*, vol. 2, see p. 171; Heaviside, *Electromagnetic Theory*, vol. 1, 135.

59. August Föppl, *Vorlesungen ueber Technische Mechanik*, Vol. 1, *Einfuehrung in die Mechanick* (1899; 2nd ed. Leipzig: B. G. Teubner, 1900), v.

60. *Ibid.*

61. De Morgan, *Trigonometry and Double Algebra*, 99.

62. *Ibid.*, iv.

63. George Boole, *The Mathematical Analysis of Logic, being an Essay towards a Calculus of Deductive Reasoning* (Cambridge: Macmillan, Barclay & Macmillan; London: George Bell, 1847), 4.

Chapter 6: Math Is Rather Flexible

1. Berkeley, *A Defence of Free-Thinking in Mathematics*, sec. XX.

2. Jean Dieudonné, *Foundations of Modern Analysis* (New York: Academic Press, 1960), 1.

3. For example, James Stewart, *Calculus: Early Transcendentals*, 4th ed. (Pacific Grove, Calif.: Brooks/Cole, 1999), A7.

4. Dieudonné, *Foundations of Modern Analysis*, 98; italics in the original.

5. Euler, article 132, *Vollständige Anleitung zur Algebra*, 79; *Élémens d'algèbre* (Lyon and Paris: Jean-Marie Bruyset, and la veuve Desaint, 1774), 98.

6. For example, Morris Kline, *Mathematics: The Loss of Certainty*, 121; Nahin, *Imaginary Tale*, 12.

7. Euler, articles 122, 150, *Algebra*, 72, 88–89.

8. Euler, article 147, *Algebra*, 87.

9. Leonhard Euler, "Recherches sur les Racines Imaginaires des Equations" (1749), *Memoires de l'Académie des Sciences de Berlin* 5 (1751), 256–257, 224–225; reprinted in Leonhardi Euleri, *Opera Omnia*, 1st ser. *Opera Mathematica*, vol. 6 (Lipsiae et Berolini: B. G. Teubneri, 1921), 113, 80–81.

10. Etienne Bézout, *Cours de Mathématiques a l'Usage du Corps Royal de l'Artillerie*, part 2, *Contenant l'Algèbre et l'application de l'Algèbre a la Géométrie*

(Paris: P. D. Pierres, Imprimeur du Roi, 1781), 95; Sylvestre-François Lacroix, *Élémens d'Algèbre, a l'usage de l'École des Quatre-Nations*, 11th ed. (Paris: Courcier, 1815), 233, 239–240; Bonnycastle, *A Treatise on Algebra*, 23–24.

11. Peacock, *Treatise on Algebra*, 449.

12. For example, Isaac Todhunter, *Algebra for the Use of Colleges and Schools*, 7th ed., rev. (London: Macmillan and Co., 1875), 213–214; Charles Smith, *A Treatise on Algebra*, 2nd ed. (London: Macmillan, 1875), 221.

13. Gottlob Frege, *Die Grundlagen der Arithmetik* (1884), trans. by J. L. Austin as *The Foundations of Arithmetic*, 2nd ed., rev. (Evanston, Ill.: Northwestern University Press, 1996), sec. 97, p. 108.

14. Etienne Bézout, *Cours de Mathématiques a l'usage des Gardes du Pavillon, de la Marine, et des Éleves de l'École Politechnique*, part 3, *Contenant l'Algèbre et l'application de cette science à l'Arithmétique et la Géométrie*, new ed., reviewed and corrected by J. G. Garnier (Paris: Courcier, Imprimeur Libraire pour les Mathématiques, 1802), 132. Lacroix, *Élémens d'Algèbre*, 239–240.

15. Frege, *Foundations of Arithmetic*, sec. 100, p. 110.

16. Isaac Todhunter, *Algebra for the Use of Colleges and Schools*, 2nd ed., rev. (London: Macmillan and Co., 1860), 31.

17. De Morgan, *Trigonometry and Double Algebra*, 93; italics in the original.

18. R. C. H. Tanner, "The Alien Realm of the Minus: Deviatory Mathematics in Cardano's Writings," *Annals of Science* 37 (1980), 159–178.

19. R. C. H. Tanner, "The Ordered Regiment of the Minus Sign: Off-beat Mathematics in Harriot's Manuscripts," *Annals of Science* 37 (1980), 127–158.

20. Hamilton, "Theory of Conjugate Functions, or Algebraic Couples," 293–422.

21. For example, Mahavira, *Ganita-Sara-Sangraha* (ca. 850), in M. Rangacarya, *The Ganita-Sara-Sangraha* (Madras: Government Press, 1912), 7. Or, for a much later statement, see Simpson, *Treatise of Algebra*, 272.

22. David Eugene Smith, *History of Mathematics*, vol. 1 (1925), reissued (New York: Dover Publications, 1953), 257; Cajori, *A History of Mathematical Notations*, vol. 1, 258; Boyer, *A History of Mathematics*, 242.

23. Tanner, "Ordered Regiment of the Minus," 127.

24. Banesh Hoffmann, *About Vectors* (Englewood Cliffs, N.J.: Prentice-Hall, 1966); reissued (New York: Dover Publications, 1975), 62; italics in the original.

25. *Ibid.*, 10; italics in the original.

26. *Ibid.*, 14.

27. "Art. II. Memoire sur les Quantités Imaginaires. Par M. Buëe," 308.

28. *Ibid.*, 317; italics in the original.

Chapter 7: Making a Meaningful Math

1. *Encyclopædia Britannica*, 80.

2. Farrar, *Introduction to the Elements of Algebra*, 5.

3. Bourdon, *Éléments d'Algèbre*, 73–74.

4. W. R. Hamilton, *Elements of Quaternions*, vol. 1 (1866, 2nd ed., ed. Charles Jasper Joly, London and New York: Longmans, Green and Co., 1901), 111; and see also p. 170; italics in the original.

5. Bertrand Russell, article 219 in *Principles of Mathematics* (1903, 2nd ed., New York: W. W. Norton & Company, 1938), 229.

6. De Morgan, *Study and Difficulties*, 105–109.

7. William Kingdon Clifford, *The Common Sense of the Exact Sciences*, ed. Karl Pearson (1885, newly ed. James R. Newman, New York: Alfred A. Knopf, 1946), 30–31.

8. Davis, *Introduction to the Logic of Algebra*, 47.

9. Föppl, *Einfuehrung in die Mechanick*, v–vi.

10. Morris Kline, "Arithmetics and their Algebras" in *Mathematics for Liberal Arts* (Reading, Mass.: Addison-Wesley, 1967); reprinted as *Mathematics for the Nonmathematician* (New York: Dover Publications, 1985), 481–482.

11. Albert Einstein, "Geometrie und Erfahrung," address to the Prussian Academy of Sciences in Berlin, 27 January 1921, expanded version in Einstein, *Sidelights on Relativity*; trans. G. B. Jeffery and W. Perrett as "Geometry and Experience" (London: Methuen & Co.; New York: E. P. Dutton & Co., 1922; reissued New York: Dover Publications, 1983); quotation on p. 28.

12. Boole, *Investigation of the Laws of Thought*, 67–68.

13. Dedekind, *Was sind und was sollen Zahlen*, preface.

14. Hoffmann, *About Vectors*, 33.

15. For example, see Henri Poincaré, *La Science et l'Hypothèse* (1902); trans. George Bruce Halsted in *Foundations of Science* (New York: The Science Press, 1913), 39, 64–65.

further reading

The following books were written for general audiences and may well serve as a continued introduction to some of the subjects discussed in the present work.

William Kingdon Clifford, *The Common Sense of the Exact Sciences*, edited and with a preface by Karl Pearson (1885), newly edited with an introduction by James R. Newman, preface by Bertrand Russell (New York: Alfred A. Knopf, 1946).

Tobias Dantzig, *Number, The Language of Science. A Critical Survey written for the Cultured Non-Mathematician* (1930), Fourth Edition (New York: The MacMillan Company, 1954).

Banesh Hoffmann, *About Vectors* (Englewood Cliffs, N.J.: Prentice-Hall, 1966), reissued (New York: Dover Publications, 1975).

Morris Kline, *Mathematics: The Loss of Certainty* (1980), reissued (New York: Oxford University Press, 1982).

Helena M. Pycior, *Symbols, Impossible Numbers, and Geometric Entanglements. British Algebra through the Commentaries on Newton's Universal Arithmetick* (Cambridge: Cambridge University Press, 1997).

Much of the research for *Negative Math* was carried out at the Libraries of the Smithsonian Institution, especially the Dibner Library of the History of Science and Technology, and also at the Library of Congress, the Libraries of the University of Minnesota, the Libraries of Harvard University, and the Burndy Library at the Massachusetts Institute of Technology. Accordingly, most of the works cited in this book can be found in such collections.

Furthermore, and finally, the following bibliography consists of a selection of books and articles dealing especially with the relationship between physical experience and mathematics, and more generally with the history, philosophy, and pedagogy of mathematics.

Accardi, L., and A. Fedullo. "On the Statistical Meaning of Complex Numbers in Quantum Mechanics," *Lettere al Nuovo Cimento* 34 (1982), 161–172.

Anglin, W. S. *Mathematics, A Concise History and Philosophy* (New York: Springer-Verlag, 1994).

Archimedes. *Geometrical Solutions Derived from Mechanics*, also known as "The Method," translated by J. L. Heiberg (Chicago: Open Court, 1909).

Ascher, M. *Ethnomathematics: A Multicultural View of Mathematical Ideas* (San Francisco: Brooks/Cole, 1991).

Aspray, W., and P. Kitcher, eds. *History and Philosophy of Modern Mathematics*, Minnesota Studies in the Philosophy of Science vol. XI (Minneapolis: University of Minnesota Press, 1988).

Azzouni, J. *Metaphysical Myths, Mathematical Practice: The Ontology and Epistemology of the Exact Sciences* (Cambridge: Cambridge University Press, 1994).

Azzouni, J. "Applying Mathematics: An Attempt to Design a Philosophical Problem," *The Monist* 83 (2000), 209–227.

Azzouni, J. *Knowledge and Reference in Empirical Science* (London: Routledge, 2000).

Bauersfeld, H. "'Language Games' in Mathematics Classrooms: Their Function and Their Effects." In *The Emergence of Mathematical Meaning: Interaction in Classroom Cultures*, edited by P. Cobb and H. Bauersfeld (Hillsdale, N.J.: Lawrence Erlbaum Associates, 1995), 271–292.

Birkhoff, G. "The Mathematical Nature of Physical Theories," *American Scientist* 31 (1943), 281–310.

Bochner, S. *The Role of Mathematics in the Rise of Science* (Princeton, N.J.: Princeton University Press, 1966).

Bos, H. *"Queen and Servant": The Role of Mathematics in the Development of the Sciences*, Lectures in the History of Mathematics vol. 7 (Providence, R.I.: American Mathematical Society; London: London Mathematical Society, 1993).

Boyer, C. "Fundamental Steps in the Development of Numeration," *Isis* 35 (1944), 157–158.

Boyer, C. *History of Analytic Geometry* (New York: Scripta Mathematica, 1956).

Boyer, C. *The History of the Calculus and Its Conceptual Development* (New York: Dover, 1959).

Brainerd, C. J. "The Origins of Number Concepts," *Scientific American* (March 1973), 101–109.

Brower, L.E.J. *Brower's Cambridge Lectures on Intuitionism*, edited by D. Van Dalen (New York: Cambridge University Press, 1981).

Browder, F. "Does Pure Mathematics have a Relation to the Sciences?" *American Scientist* **64** (1976), 542.

Brown, S. I. "Thinking Like a Mathematician: A Problematic Perspective," *For the Learning of Mathematics* **17**:2 (1997), 35–38.

Bunge, M. *Intuition and Science* (Englewood Cliffs, N.J.: Prentice-Hall, 1962).

Burington, A. S. "On the Nature of Applied Mathematics," *American Mathematical Monthly* **56** (1949), 221–241.

Calder, A. "Constructive Mathematics," *Scientific American* **241** (Oct. 1979), 134–143.

Castonguay, C. *Meaning and Existence in Mathematics* (New York: Springer-Verlag, 1972).

Chihara, C. *Ontology and the Vicious Circle Principle* (Ithaca, N.Y.: Cornell University Press, 1973).

Conant, L. L. *The Number Concept* (London: MacMillan, 1923).

Courant, R., and H. Robbins. *What Is Mathematics?* (New York: Oxford University Press, 1948).

Crowe, M. J. *A History of Vector Analysis* (New York: Dover, 1967).

Crowe, M. J. "Ten 'Laws' Concerning Patterns of Change in the History of Mathematics," *Historia Mathematica* **2** (1975), 161–166.

Crowe, M. J. "Ten Misconceptions About Mathematics and Its History," in Aspray and Kitcher (1988).

Crump, T. *The Anthropology of Numbers* (Cambridge: Cambridge University Press, 1990).

Davis, C. "Materialist Mathematics," *Boston Studies in the Philosophy of Science*, vol. 15 (Dordrecht: Reidel, 1974), 37–66.

Davis, P. J., "Mathematics by Fiat?" *The Two-Year College Mathematics Journal* **11** (Sept. 1980), 255–263.

Davis, P. J., and R. Hersh. *The Mathematical Experience* (Boston: Birkhauser, 1981).

de Broglie, L. "The Role of Mathematics in the Development of Contemporary Theoretical Physics." In *Great Currents of Mathematical Thought*, vol. 2, edited by F. Le Lionnais (New York: Dover, 1971), 78–93.

Dedekind, J.W.R. *Essays on the Theory of Numbers* (LaSalle, Ill.: Open Court, 1901).

De Morgan, A. *On the Study and Difficulties of Mathematics* (London: Open Court, 1915).

Doncel, M., et al. *Symmetries in Physics, 1600–1980* (Barcelona: Universidad Autónoma de Barcelona, 1987).

Dummett, M. *Elements of Intuitionism* (Oxford: Clarendon Press, 1977).

Dummett, M. "What is Mathematics About?" In *Mathematics and Mind*, edited by A. George (Oxford: Oxford University Press, 1994), 11–26.

Dyson, F. J. "Mathematics in the Physical Sciences," *Scientific American* **211** (Sept. 1964), 129–146; also in Committee on the Support of Research in the

Mathematical Sciences (COSRIMS) of the National Research Council, ed., *The Mathematical Sciences* (Cambridge, Mass.: MIT Press, 1969).

Epstein, D., and S. Levy. "Experimentation and Proof in Mathematics," *Notices of the American Mathematical Society* 42 (1995), 670–674.

Ernest, P. *The Philosophy of Mathematics Education* (New York: Falmer, 1991).

Fauvel, J., and J. van Maanen. "The Role of History of Mathematics in the Teaching and Learning of Mathematics: Discussion Document for an ICMI Study," *Educational Studies in Mathematics* 34:3 (1997), 255–259.

Field, H. *Science without Numbers* (Princeton, N.J.: Princeton University Press, 1980).

Frege, G. *The Foundations of Arithmetic* (Evanston, Ill.: Northwestern University Press, 1980).

Freundenthal, H. *Mathematics as an Educational Task* (Dordrecht: Reidel, 1973).

Furth, M. *The Basic Laws of Arithmetic* (Berkeley: University of California Press, 1964).

Goodman, N. "Mathematics as an Objective Science," *American Mathematical Monthly* 81 (1974), 354–365.

Goodman, N. "Mathematics as an Objective Science," *American Mathematical Monthly* 86 (1979), 540–541.

Hale, W. T. "UICSM's Decade of Experimentation," *The Mathematics Teacher* 54 (1961), 613–619.

Halstead, G. B. *On the Foundation and Technic of Arithmetic* (London: Open Court, 1912).

Hamming, R. W. "The Unreasonable Effectiveness of Mathematics," *American Mathematical Monthly* 87 (1980), 81–90.

Hardy, G. H. *A Mathematician's Apology* (New York: Cambridge University Press, 1940).

Hatfield, G. *The Natural and the Normative* (Cambridge, Mass.: MIT Press, 1990).

Hatton, J.L.S. *The Theory of the Imaginary in Geometry, together with the Trigonometry of the Imaginary* (Cambridge: Cambridge University Press, 1920).

Heath, T. L. *Mathematics in Aristotle* (Oxford: Oxford University Press, 1949).

Heath, T. L. *A History of Greek Mathematics* (New York: Dover, 1981).

Hefendehl-Hebeker, L. "Negative Numbers: Obstacles in their Evolution from Intuitive to Intellectual Constructs," *For the Learning of Mathematics* 11:1 (1991), 26–32.

Helmholtz, H. *Epistemological Writings* (Boston: Reidel, 1977).

Helmholtz, H. *Counting and Measuring* (New York: D. Van Nostrand Company, 1980).

Hempel, C. G. "Geometry and Empirical Science," *American Mathematical Monthly* 52 (1945), 7–17.

Henrici, P. "Reflections of a Teacher of Applied Mathematics," *Quarterly of Applied Mathematics* **30** (1972), 31–39.

Hersh, R. "Proving is Convincing and Explaining," *Educational Studies in Mathematics* **24** (1993), 389–399.

Hersh, R. *What is Mathematics, Really?* (New York: Oxford University Press, 1997).

Hestenes, D. "Spacetime Physics with Geometric Algebra," *American Journal of Physics* **71**:7 (Jul. 2003), 691–714.

Heyl, P. R. "The Skeptical Physicist," *Scientific Monthly* **46** (Mar. 1938), 225–229.

Hoffmann, B. *About Vectors* (New York: Dover, 1975).

Hughes, R.I.G. *The Structure and Interpretation of Quantum Mechanics* (Cambridge, Mass.: Harvard University Press, 1989).

Hume, D. *An Inquiry Concerning Human Understanding* (Indianapolis: Bobbs-Merrill, 1995).

Iliev, L. "Mathematics as the Science of Models," *Russian Mathematical Surveys* **27** (1972), 181–189.

Irvine, A. D., ed. *Physicalism in Mathematics* (Dordrecht: Kluwer, 1990).

Isaacson, D. "Mathematical Intuition and Objectivity." In *Mathematics and the Mind*, edited by A. George (New York: Oxford University Press, 1994).

Jesseph, D. M. *Berkeley's Philosophy of Mathematics* (Chicago: University of Chicago Press, 1993).

Johnson, M. *The Body in the Mind: The Bodily Basis of Meaning, Reason and Imagination* (Chicago: University of Chicago Press, 1987).

Kamii, C., and M. A. Warrington. "Division with Fractions: A Piagetian, Constructivist Approach," *Hiroshima Journal of Mathematics Education* **3** (1995), 53–62.

Kitchener, R. F. *Piaget's Theory of Knowledge* (New Haven, Conn.: Yale University Press, 1986).

Kitcher, P. "Mathematical Naturalism," in Aspray and Kitcher (1988).

Klein, F. *Development of Mathematics in the 19th Century* (Brookline, Mass.: Math Science Press, 1979).

Klein, J. A. *Greek Mathematical Thought and the Origin of Algebra* (Cambridge, Mass.: MIT Press, 1968).

Kline, M. *Mathematics and the Physical World* (New York: Doubleday, 1963).

Kline, M. *Mathematical Thought from Ancient to Modern Times* (New York: Oxford University Press, 1972).

Lampert, M. "When the Problem Is Not the Question and the Solution Is Not the Answer: Mathematical Knowing and Teaching," *American Educational Research Journal* **27**:1 (1990), 29–64.

Ma, L. *Knowing and Teaching Elementary Mathematics: Teachers' Understanding of Fundamental Mathematics in China and the United States* (Mahwah, N.J.: Lawrence Erlbaum Associates, 1999).

Mack, N. K. "Learning Fractions with Understanding: Building on Informal Knowledge," *Journal for Research in Mathematics Education* **21** (1990), 16–32.

Mackey, G. W. *Mathematical Foundations of Quantum Mechanics* (New York: Benjamin, 1963).

Maddy, P. *Realism in Mathematics* (New York: Oxford University Press, 1984).

Madell, R. "Children's Natural Processes," *Arithmetic Teacher* **32** (1985), 20–22.

Manin, Y. I. *Mathematics and Physics* (Boston: Birkhauser, 1981).

Margenau, H. "Philosophical Problems Concerning the Meaning of Measurement in Physics," *Philosophy of Science* **25** (1958), 23–33.

Maxwell, J. C. "On the Mathematical Classification of Physical Quantities," *Proceedings of the London Mathematical Society* **3** (1871), 224–232. Reprinted in W. D. Niven, ed., *The Scientific Papers of James Clerk Maxwell*, vol. 2 (New York: Dover, 1965), 257–266.

McAulay, A. *Utility of Quaternions in Physics* (London: MacMillan and Co., 1893).

Mehrtens, H. T. "T. S. Kuhn's Theories and Mathematics," *Historia Mathematica* **3** (1976), 297–320.

Mickens, R. E., ed. *Mathematics and Science* (Singapore: World Scientific, 1990).

Mill, J. S. *A System of Logic, Ratiocinative and Inductive*, 8th ed. (New York: Harper & Brothers, 1874).

Moon, P., and D. E. Spencer. *Vectors* (Princeton, N.J.: D. Van Nostrand Company, 1965).

Nagel, E. "Impossible Numbers: A Chapter in the History of Modern Logic," *Studies in the History of Ideas*, vol. 3 (New York: Columbia University Press, 1935).

Nahin, P. J. *An Imaginary Tale: The Story of $\sqrt{-1}$* (Princeton, N.J.: Princeton University Press, 1998).

Nelsen, R. B. *Proofs Without Words. Exercises in Visual Thinking* (Washington, D.C.: Mathematical Association of America, 1993).

Park, J. L. "The Logic of Noncommutability of Quantum Mechanical Operators and Its Empirical Consequences." In *Perspectives in Quantum Theory: Essays in Honor of Alfred Landé*, edited by W. Yourgrau and A. van der Merwe (Cambridge, Mass.: MIT Press, 1971), 37–70.

Pearson, K. *The Grammar of Science*, 3rd ed. (London: Adam and Charles Black, 1911).

Peirce, C. S. "The Essence of Mathematics." In *The World of Mathematics* vol. 3, edited by J. R. Newman (New York: Simon and Schuster, 1956), 1773–1783.

Peirce, C. S. *The New Elements of Mathematics* (The Hague: Mouton, 1976).

Penrose, R. "The Geometry of the Universe." In *Mathematics Today: Twelve*

Informal Essays, edited by L. A. Steen (New York: Springer-Verlag, 1978), 83–125.

Piaget, J. "How Children Form Mathematical Concepts," *Scientific American* **189** (1953), 74–79.

Piaget, J. *The Child's Conception of Number* (New York: Norton, 1965).

Piaget, J., B. Inhelder, and A. Szeminka. *The Child's Conception of Geometry* (New York: Basic Books, 1980).

Pimm, D. *Symbols and Meanings in School Mathematics* (London: Routledge, 1995).

Poincaré, H. *The Foundations of Science* (New York: Science Press, 1913).

Poincaré, H. *Mathematics and Science; Last Essays* (New York: Dover, 1963).

Pólya, G. *Mathematics and Plausible Reasoning* (Princeton, N.J.: Princeton University Press, 1954).

Pour-El, M. B., and J. I. Richards. *Computability in Analysis and Physics* (Berlin: Springer-Verlag, 1989).

Putnam, H. "What Is Mathematical Truth?" In *Mathematics, Matter and Method* (London: Cambridge University Press, 1975), 67–78.

Putnam, H. *Representation and Reality* (Cambridge, Mass.: MIT Press, 1988).

Pycior, H. M. "Peirce at the Intersection of Mathematics and Philosophy." In *Peirce and Contemporary Thought: Philosophical Inquiries*, edited by K. L. Ketner (New York: Fordham University Press, 1995), 132–145.

Radford, L. "Signs and Meanings in Students' Emergent Algebraic Thinking: A Semiotic Analysis," *Educational Studies in Mathematics* **42** (2000), 237–268.

Radford, L. "Students' Processes of Symbolizing in Algebra. A Semiotic Analysis of the Production of Signs in Generalizing Tasks." In *Proceedings of the 24th Conference of the International Group for the Psychology of Mathematics Education (PME-24)*, vol. 4, edited by T. Nakahara and M. Koyama (Hiroshima: Hiroshima University Press, 2000), 81–88.

Reed, M., and B. Simon. *Methods of Modern Mathematical Physics* (New York: Academic Press, 1975).

Resnick, M. D. *Mathematics as a Science of Patterns* (Oxford: Oxford University Press, 1997).

Richards, J. *Mathematical Visions: The Pursuit of Geometry in Victorian England* (San Diego, Calif.: Academic Press, 1988).

Riemann, B. "On the Hypotheses which Lie at the Bases of Geometry." Translated by W. K. Clifford, *Nature* **8** (1873), 14–17, 36–37.

Rosefeld, B. S. *A History of Non-Euclidean Geometry* (New York: Springer-Verlag, 1988).

Rotman, B. *Signifying Nothing: The Semiotics of Zero* (New York: St. Martin's Press, 1987).

Rotman, B. "Toward a Semiotics of Mathematics," *Semiotica* **72** (1988), 1–35.

Ruelle, D. "Is Our Mathematics Natural? The Case of Equilibrium Statistical Mechanics," *Bulletin of the American Mathematical Society* 19 (1988), 259–268.

Scriba, C. J. *The Concept of Number* (Zurich: B. Mannheim, 1986).

Seidenberg, A. "The Ritual Origin of Counting," *Archive for History of the Exact Sciences* 2 (1962), 1–40.

Sfard, A. "Operational Origins of Mathematical Objects and the Quandary of Reification—The Case of Function." In *The Concept of Function*, MAA Notes, vol. 25, edited by E. Dubinsky and G. Harel (Washington, D.C.: Mathematical Association of America, 1992), 59–84.

Sfard, A. "On Reform Movement and the Limits of Mathematical Discourse," *Mathematical Thinking and Learning* 2 (2000), 157–189.

Sfard, A., and L. Linchevski. "The Gains and Pitfalls of Reification—the Case of Algebra," *Educational Studies in Mathematics* 26 (1994), 191–228.

Shapiro, S. "Mathematics and Reality," *Philosophy of Science* 50 (1984), 523–548.

Shenker, O. "Fractal Geometry is Not the Geometry of Nature," *Studies in the History and Philosophy of Science* 25 (1994), 967–981.

Sherry, D. "The Logic of Impossible Quantities," *Studies in the History and Philosophy of Science* 22 (1991), 37–62.

Sierpinska, A. "Mathematics 'in Context,' 'Pure' or 'with Applications'?" *For the Learning of Mathematics* 16:2 (1995), 2–15.

Smorynski, C. "Mathematics as a Cultural System," *Mathematical Intelligencer* 5 (1983), 9–15.

Snapper, E. "What Is Mathematics?" *American Mathematical Monthly* 86 (1979), 551–557.

Snapper, E. "What Do We Do When We Do Mathematics?" *Mathematical Intelligencer* 10 (1988), 53–58.

Sossinsky, A. *Knots: Mathematics with a Twist* (Cambridge, Mass.: Harvard University Press, 2002).

Steen, L. "The Science of Patterns," *Science* 240 (1988), 611–616.

Stein, S. K. *Mathematics, the Man-Made Universe* (San Francisco: Freeman, 1963).

Stein, S. K. "Where Have All the Reforms Gone?" In *Strength in Numbers: Discovering the Joy and Power of Mathematics in Everyday Life* (New York: John Wiley & Sons, 1996), 74–93.

Stein, S. K. "Why Negative Times Negative Is Positive." *Ibid.*, 189–193.

Steiner, M. *Mathematical Knowledge* (Ithaca, N.Y.: Cornell University Press, 1975).

Steiner, M. *The Applicability of Mathematics as a Philosophical Problem* (Cambridge, Mass.: Harvard University Press, 1998).

Sternberg, S. *Group Theory and Physics* (Cambridge: Cambridge University Press, 1994).

Tait, P. G. "On the Importance of Quaternions in Physics," *Philosophical Magazine* (1890). Reprinted in *Scientific Papers*, vol. 2 (Cambridge: Cambridge University Press, 1900), doc. 97.

Thom, R. "Modern Mathematics: An Educational and Philosophical Error?" *American Scientist*, **59** (1971), 695–699.

Thomaidis, Y. "Aspects of Negative Numbers in the Early 17th Century," *Science and Education* **2** (1993), 74.

Thomas, R. "Proto-Mathematics and/or Real Mathematics," *For the Learning of Mathematics* **16**:2 (1996), 11–18.

von Neumann, J. "The Mathematician." In *The Works of the Mind*, edited by R. B. Heywood (Chicago: University of Chicago Press, 1947), 180–196.

von Neumann, J. *Mathematical Foundations of Quantum Mechanics*, translated by R. T. Beyer (Princeton: Princeton University Press, 1955).

Wang, H. "Toward Mechanical Mathematics," International Business Machines Corporation, 1960; reprinted in *Automation of Reasoning*, edited by J. Siekmann and G. Wrightson (Berlin: Springer-Verlag, 1983), 229–266.

Weinberg, S. "Lecture on the Applicability of Mathematics," *Notices of the American Mathematical Society* **33**.5 (Oct. 1986).

Weyl, H. *Symmetry* (Princeton, N.J.: Princeton University Press, 1952).

White, M. J. *The Continuous and the Discrete* (New York: Oxford University Press, 1992).

Whitney, H. "The Mathematics of Physical Quantities," *American Mathematical Monthly* **75** (1968), 115, 227.

Wiener, N. *I Am a Mathematician* (Garden City, N.Y.: Doubleday, 1956).

Wigner, E. "The Unreasonable Effectiveness of Mathematics in the Natural Sciences," *Communications on Pure and Applied Mathematics* **13** (1960), 1–14; also in *Symmetries and Reflections* (1967), 222–237.

Wigner, E. *Symmetries and Reflections* (Bloomington: Indiana University Press, 1967).

Wilson, E. B. *Vector Analysis: A Text-Book for the Use of Students of Mathematics and Physics*, based on the lectures of J. W. Gibbs (New Haven, Conn.: Yale University Press, 1901).

Wilder, R. L. *Evolution of Mathematical Concepts: An Elementary Study* (New York: Wiley, 1968).

Wilder, R. L. *Mathematics as a Cultural System* (New York: Pergamon, 1981).

Windred, G. "History of the Theory of Imaginary and Complex Quantities," *The Mathematical Gazette* **14** (1929), 533–541.

Wittgenstein, L. *Lectures on the Foundations of Mathematics* (Ithaca, N.Y.: Cornell University Press, 1976).

Wittgenstein, L. *Remarks on the Foundations of Mathematics* (Cambridge, Mass.: MIT Press, 1983).

Wu, H. "The Mathematics Education Reform: Why You Should Be Concerned and What You Can Do," *American Mathematical Monthly* **104** (1997), 946–954.

Yanin, Y. *Mathematics and Physics* (Boston: Birkhauser, 1983).

Zee, A. "The Effectiveness of Mathematics in Fundamental Physics." In *Mathematics and Science*, edited by R. E. Mickens (Singapore: World Scientific, 1990), 307–323.

acknowledgments

The following individuals kindly read one or another version of the manuscript and gave me helpful comments: David Lindsay Roberts, Olivia Walling, Michael Brian Schiffer, Abigail J. Lustig, Ronald Martínez Cuevas, Elizabeth Cavicchi, John L. Heilbron, Simon W. Albright, Joan Richards, Ronald Anderson. I thank them all for valuable feedback and suggestions. Earlier, Roger H. Stuewer had revised parts of the historical material, which appeared in my dissertation, and he painstakingly taught me how to write history clearly. I apologize for not always following his good advice. I also thank the reviewers at Princeton University Press for their thoughtful, sometimes critical, and entirely helpful comments and suggestions. I especially thank my editor, Vickie Kearn, for taking an interest in this project and for giving me her friendly and efficient support. I also appreciate the work done by my copyeditor Jennifer Slater, and her sharp blue pencil. Likewise, I thank Meera Vaidyanathan, Alycia M. Somers, Dimitri Karetnikov, and the other individuals who contributed in the review and production of this book at Princeton University Press. I also warmly thank the following individuals for kindly helping me in various ways during the writing process: Ronald L. Brashear, Lindsey Denise Cameron, Bonnie Edwards, Michael G. Fisher, Alison Jones, Tien-Yi Lee, Anahid Mikirditsian, my mother Lillian Montalvo Conde, Roberto José Ortíz, Elizabeth Paris, Kim Pfloker, Alberto Quintero, Justine Ellen Roberts, Erwin N. Hiebert, George E. Smith, Bonnie Sousa, John Stachel, Kirsten van der Veen, Daria Wingreen, Emily W. Allen, Kristin Von Kundra. I also thank the helpful staffs at the following libraries: the Dibner Library for the History of Science and Technology at the Smithsonian Institution, the Burndy Library at the Dibner Institute for the History of Science and Technology, the Widener Library and the Houghton Library of Harvard University. Furthermore, I thank the historians, mathematicians, and writers whose works I found so valuable and instructive during this project.